Rivertown

Urban and Industrial Environments
Series editor: Robert Gottlieb, Henry R. Luce Professor of Urban and Environmental Policy, Occidental College

For a complete list of books published in this series, please see the back of the book.

Rivertown

Rethinking Urban Rivers

edited by
Paul Stanton Kibel

The MIT Press
Cambridge, Massachusetts
London, England

For information about special quantity discounts, please e-mail <special_sales@ mitpress.mit.edu>.

This book was set in Sabon by SPi Publisher Services.

Printed on recycled paper and bound in the United States of America.

Library of Congress Cataloging-in-Publication Data

Library of Congress Cataloging-in-Publication Data
Rivertown : rethinking urban rivers / edited by Paul Stanton Kibel.
    p. cm.—(Urban and industrial environments)
  Includes bibliographical references and index.
  ISBN: 978-0-262-11307-6 (hardcover : alk. paper)
  ISBN: 978-0-262-61219-7 (pbk. : alk. paper)
  1. Urban renewal—United States. 2. Waterfronts—United States. I. Kibel, Paul Stanton.

HT175.R58 2007
307.3'4160973—dc22

                                                                    2006029837

10   9   8   7   6   5   4   3   2   1

For my daughter Helen as she heads downstream
—PSK

Contents

Rivertown

1

Bankside Urban: An Introduction

Paul Stanton Kibel

Renewal Rethought

The impetus for this book was the pervasiveness of a tangible change in our nation's urban riverfronts. Riverside city lands previously devoted to heavy industry and shipping were being converted to housing, open space, and parks. Concrete bulkheads and embankments were being removed and replaced with restored wetlands, marshes, and beaches. Warehouses, highway overpasses, and other structures that formerly crowded the waterline were being torn down or set back to allow greater public access. Urban rivers, formerly the open sewers of our cities, were increasingly identified as resources needed to support waterfowl, fisheries, and canoeing.

This change was made possible by multiple factors that rendered lands along city rivers less economically viable for former industrial or maritime uses. These factors included shifts in international travel from passenger ship to aircraft, shifts in domestic cargo transport from ship to truck and rail, a change from port-based commercial fishing to deep-sea trawlers, maritime freight containerization that could not be handled at cramped downtown docks lacking modern loading and off-loading facilities, the decline of heavy industries located on urban riverfronts and the relocation of such industries to suburban or rural sites or abroad, and the demand of riverside community residents for increased public parkland.[1]

The recent changes on the riverfront, however, carry echoes of a previous period of transformation in American cities—the urban renewal of the 1940s, 1950s, and 1960s. These echoes can be found not only in many of the physical riverside structures now targeted for removal but also in debates over the process by which urban bankside land-use decisions should be made.

Slum clearance was the term commonly used to describe the urban-renewal policies implemented following the passage of the federal Housing Acts of 1937 and 1949 and the federal Highway Act of 1956. These laws worked in close tandem, expanding on policies first put into place by the Federal Housing Administration (FHA) in 1934.[2]

The Federal Housing Administration adopted criteria that denied mortgage insurance to most older buildings in urban neighborhoods with high-minority, low-income residents.[3] For instance, under FHA standards 50 percent of Detroit and 33 percent of Chicago were blacked out as ineligible for mortgage insurance.[4] Without such mortgage insurance, people wishing to purchase or restore such buildings could not obtain financing, which led to a decline in the value and condition of such properties in the late 1930s and 1940s. These same neighborhoods were then declared slums under the 1937 and 1949 Housing Acts, and the deteriorating buildings (the homes and business of the people in these neighborhoods) were acquired cheaply through eminent domain and demolished. The cleared lands were then often used for multistory public housing or for freeways built with funds provided under the 1956 Highway Act. This, in abbreviated form, was the paradigm for slum clearance and urban renewal in the midtwentieth century.

It was a paradigm, however, in which the residents living in the areas subject to urban renewal often ended up the victims rather than the beneficiaries of this clearance. As Richard Moe and Carter Wilkie note in their 1997 book, *Changing Places: Rebuilding Community in the Age of Sprawl*, "Federal home mortgage insurance policies helped to guarantee the conversion of once sound urban areas to urban slums. In existing neighborhoods, the effect was further declining property values, which means fewer taxes to the city treasury, which stretched the city's budget and led to a decline in services, which encouraged more residents to move out only to be replaced by residents of lower means."[5] Moe and Wilkie continue: "By clearing slums from the urban landscape, such reasoning went, the conditions that produced the slums in the first place could be eliminated. . . . All too frequently, the results were something else: monstrous housing projects of faulty design, poorly planned neighborhoods that were seldom integrated with surrounding areas, and, in most cases, the displacement of residents without the provision of alternative housing."[6] Their conclusion: "Though urban renewal gave some fading locations new leases on economic life, many others were left in ruins, typically those with the poorest and least politically connected constituencies."[7]

Even urban historian Jon C. Teaford, generally considered a defender of and apologist for slum-clearance policies, refers to highway construction, urban renewal, and redlining by financial institutions in midcentury America as the "trinity of evil," noting: "The first and second would kill the neighborhoods through a quick blow from the bulldozer. The third would slowly cut off the financial lifeblood of the community by denying mortgages and home improvement and business loans in neighborhoods that bankers deemed undesirable."[8] Teaford concedes, "In the early 1960s renewal was already beginning to lose its appeal; by the 1970s it had become a dirty word, and because of this stigma urban renewal agencies were changing their names to departments of community development."[9]

The previous experiences with urban renewal and slum clearance are a necessary backdrop for understanding the current discussion regarding urban river restoration. Many of the freeways and housing constructed in the 1940s, 1950s, and 1960s are along city rivers. On the one hand, therefore, today's urban river-restoration efforts offer an opportunity to address the mistakes and consequences of the earlier urban-renewal period by removing certain structures and giving the local community a meaningful role in land-use decisions going forward. On the other hand, however, current river-restoration efforts may also lead to an unfortunate repeat of urban renewal, by once again upending, dislocating, and politically marginalizing current residents in the name of redevelopment. Although the official language of most urban riverfront-restoration plans is now couched in terms of "community development," at times the substance of these plans seems like blight-removal redux. The new blight may be public housing and industry, and the replacement for this new blight may be gentrified office, retail, and residential space, but for the residents and workers in these blighted areas, "community development" can still seem like yet another program to move them out.

The current debates over the use of urban riverside lands therefore raise questions that are of particular concern in the post–urban-renewal era. If parkland and open space are going to be created, who will be the primary users and beneficiaries of these new resources? Will new riverfront proposals come from within the community where these lands are located or from developers outside the community? What role will governmental agencies and policies play in this process?

These dynamics are evident in riverfront cities across the United States. Take, for example, the 2001 mayoral race in Los Angeles. During this campaign, the candidates came together in September 2000 for a forum at Occidental College near downtown to discuss the urban environment in general and the Los Angeles River in particular. The event was broadcast on local public radio station KCRW. During the forum, the candidates spoke repeatedly on the need for residents living near the river to participate in decisions affecting riverside lands.

Mayoral candidate and Los Angeles city councilmember Joel Wachs explained his opposition to a proposed manufacturing and warehouse complex to be built on Chinatown's Cornfield property near the Los Angeles River: "I voted against it because I thought the project was bad for the city—it ignored the needs of the community—and I voted against it because the process was incredibly flawed. It is a process that left the people most affected out. . . . There was no voice for the very communities that are going to have to live there and be affected by it."[10]

Mayoral candidate and U.S. Congressman Javier Becerra concurred with Wachs, adding: "You need to reopen the process; you need to be able to have full accountability on the way the process was conducted. You also have to make sure that there's input by those who live in and around the area. Those are the folks who are going to be most impacted and they deserve to know exactly how things are being done."[11]

Mayoral candidate and speaker of the California State Assembly Antonio Villaraigosa also came out against the riverside manufacturing and warehouse project: "All of the organizations that have been working to revitalize that river should be heard and listened to. The communities in and surrounding the Cornfield need to be taken into account. . . . We need warehouses, but we don't need them in the optimum space where we can green up this city."[12]

Whether Villaraigosa (who lost the 2001 mayoral race to James Hahn but came back to defeat Hahn in 2005) and the other candidates at the Occidental College forum live up to this rhetoric remains to be seen. The pervasiveness of this rhetoric by the candidates, however, is telling in and of itself. It suggests that there is a now a widespread perception among politicians that much of the public believes something is fundamentally wrong with urban land-use redevelopment policies that run roughshod over the needs of residents for greater open space and parkland. This perception suggests that the terms of the public debate have shifted—from both environmental and participatory standpoints—since the days of urban renewal. This shift can be seen not only in the case of the Los Angeles River but also with the other urban waterways covered in this book—the Anacostia River in Washington, D.C., the Chicago River, Salt Lake's City Creek, and the Guadalupe River in San Jose.

Landscape's Undergrowth

Beyond an appreciation of the legacy and lessons of the urban renewal and slum clearance period, analysis of the changing American riverfront requires integration of different academic and professional disciplines. Such analysis involves economics, city planning, environmental and land-use law, racial politics, fisheries biology, restoration ecology, botany, real estate markets, hydrology, civil engineering, urban design, housing policy, and landscape architecture. Collectively, the essays in this book draw from these varied disciplines, but for purposes of establishing an introductory framework for the book, it makes sense to begin with the historical debates within the field of landscape architecture, which offer a particularly strategic entry point for the current debates regarding urban rivers.

Two of the seminal figures in landscape architecture in the United States are Jens Jensen and Frederick Law Olmsted. Jensen and Olmsted both held strong (and often similar) opinions about urban landscaping, but these opinions were grounded in a somewhat different set of assumptions about cities and their inhabitants.[13] A comparison of these assumptions reveals common themes that surface in the essays that follow.

Sifting through Jensen

Jens Jensen was an immigrant from Denmark whose most productive years as an urban landscaper were from 1906 to 1920, when he served as superintendent for the West Chicago Parks Commission.[14] During this period, he designed Chicago's landmark Humboldt and Columbus Parks, published his open-space study entitled *A Plan for a Greater West Park System*, and founded the conservation organization Friends of the Native

Landscape.[15] Jensen is the leading figure in the Prairie School of land-scape architecture, which, like the buildings of Prairie School architects Frank Lloyd Wright and Dwight Perkins, stressed adapting design to a site's natural setting rather than adapting a site's natural setting to accommodate design.[16] In this regard, his work has influenced many eco-logical restoration efforts on urban rivers and riverfront lands.[17]

In 1939, Jensen published a series of essays in a book entitled *Siftings*. In *Siftings*, he presents a passionate case for the Prairie School approach to landscaping: "To produce mechanical and scientific effects in plant life is foreign to the true purpose of the landscaper and to the finer feeling of mankind. . . . The skill of the landscaper lies in his ability to find the plant which needs not be maimed and distorted to fit the situation. . . . Straight lines are copies from the architect and do not belong to the landscaper. They have nothing to do with nature, of which landscaping is a part and out of which the art has grown."[18] Passages such as this provide philo-sophic sustenance for persons now working to tear down the hard edges of our city rivers and restore riverfronts to a condition that is more reflec-tive and supportive of natural ecosystems. Jensen's continuing stature is evidenced by City of Chicago's formation of the Jens Jensen Legacy Project in 2000 to document and celebrate his contributions to the fields of conservation and urban design.[19]

Siftings, however, can be a complex and at times unsettling read. For alongside his articulation of a naturalist approach to landscape architec-ture, Jensen also candidly shares his sentiments about cities and the peo-ple who populate them. After praising the admirable qualities of rural communities in the United States, Jensen states: "Contrast these expres-sions of our rural country to those of the city . . . where . . . dark corners encourage vice and dishonesty. . . . In this entanglement of masonry the growth of cunningness and trickery, conceit, and jealousy and hatred is much greater than in the free and open country."[20] In *Siftings*, Jensen con-tinues in this vein, adding that "It is from the rural country, from the farming communities and the small towns and villages, that the real American culture will eventually come"[21] and concluding that "Most of our large cities throughout the land are raging in cunning, trickery and chaos. . . . In the city man develops mob psychology and with that his freedom goes. He is no longer a single individual but a tool to be used. His whole view of his community becomes warped."[22]

Hence, while Jensen devoted much of his life's work to improving landscape design in our cities, he undertook this work with a certain disdain for urban people. What lay at the root of Jensen's antiurbanism? Several passages in *Siftings* suggest that it might be traced to Jensen's views about the relation between the effects of environment on ethnic traits and about the threat posed by foreign, nonnative species.

In his introductory essay in *Siftings*, Jensen comments: "The farther south a northern people migrate, the more degenerating are the influences of the environment, due of course to the climatic conditions which have changed their mode of living. Yet, in the mountains of Virginia, Kentucky and Tennessee the strong characteristics of a northern people have remained untouched. . . . The mountainous influence has made them more daring than their neighbors in the lowland."[23] This passage takes on an even more interesting hue when read in conjunction with Jensen's later observation in *Siftings* (in regard to disturbances caused by the introduction of the Japanese honeysuckle into the United States): "This shows the ultimate danger of transplanting plants to soil and climate foreign to their native habitat. The great destruction brought to our country through foreign importations must prove alarming to the future."[24]

While these passages from *Siftings* by themselves may not establish that Jensen's antiurban beliefs can be attributed to his efforts to preserve the "strong characteristics" of the United States' native ethnic rural stock, other Jensen writings make this connection more explicit. For instance, in 1937, when the National Socialists (Nazis) held power in Germany, Jensen published an article in the German journal *Gartenkunst* in which he explained that the gardens he designs are "in harmony with their landscape and the racial characteristics of its inhabitants. They shall express the spirit of America and therefore have to be as free of foreign character as far as possible. . . . the Latin and the Oriental crept and creeps more and more over our land, coming from the South, which is settled by Latin people, and also from other centers of mixed masses of immigrants. The Germanic character of our race, of our cities and settlements, was overgrown by foreign elements."[25] Statements like this, from a man who was himself a first-generation immigrant to the United States, make it difficult to read Jensen's denunciations against the Japanese honeysuckle without suspecting that perhaps his complaint had as much to do with the Japanese as with the honeysuckle. Such passages also raise

Figure 1.1
Chicagoans in Jensen's Humboldt Park, circa 1910. Photograph courtesy of the City of Chicago Park District.

the question of whether his founding of Friends of the Native Landscape may have been prompted in part to help preserve what Jensen perceived as the native Northern European stock of the region.

Jensen is rightfully credited for creating eloquent translations of prairie landscapes into the urban park setting and for inspiring other park designers to take their cues from the regional vegetation and terrain of the surrounding environs. As Charles E. Little, editor of the Johns Hopkins University Press's American Land Classic series, observed in his foreword to the most recent reissue of *Siftings*, "Jensen's view that we should make our designs harmonious with nature and its ecological processes was to become the preeminent theme is modern American landscape architectural practice."[26] Yet Jensen's legacy in the urban park landscaping field is complicated by the xenophobia that seems to lurk close beneath his naturalist and ecological design approach. He pursued his efforts to green our nation's cities despite his apparent dislike of the people who increasingly lived there.

Yeoman Olmsted

Frederick Law Olmsted's work predated that of Jensen. Olmsted's career as a city planner and park designer spanned roughly from 1850 to 1900. Although he designed many landmark urban parks in North America—including Prospect Park in Brooklyn and Mont Royal Park in Montreal—Olmsted's most acclaimed city park project is Central Park in New York City.

As planning for Central Park began, Olmsted made clear that he did not envision the park simply as a playground for New York City's rich. Rather, he hoped to create a surrogate wilderness experience for those who were not rich. As he explained to the New York City Park Commission in 1858: "It is one of the great purpose of the Park to supply to the hundreds of thousands of tired workers, who have no opportunity to spend their summers in the country, a specimen of God's handiwork that shall be to them, inexpensively, what a month or two in the White Mountains or the Adirondacks is, at great cost, to those in easier circumstances."[27] For Central Park, he wanted rolling hills and meadows—not flowerbeds. Olmsted fought hard for this vision, often clashing with Park Commissioners Robert Dillon and August Belmont, who pressed for inclusion of a straight, two-mile manicured carriage promenade.[28]

Through his advocacy and defense of naturalist landscapes for urban parks, Olmsted laid much of the aesthetic and ecological groundwork that underpinned Jensen's approach. However, the two men were far apart in the social beliefs that guided their park-design criteria. Unlike Jensen, who built urban parks notwithstanding his low opinion of many of persons who were likely to use them, Olmsted's interest in the welfare of these same persons helped motivate him to build urban parks. Whereas Jensen recoiled from the "mixed masses," Olmsted hoped that these masses would be the primary users and beneficiaries of his landscaped city creations.

Despite his vision for the park's primary users, Olmsted shared some of the prejudices that were prevalent in the era in which he lived. Growing up primarily among white, affluent, educated New Englanders, Olmsted had little initial contact with people outside this racial and economic class. His views evolved through his life experience, however, and although perhaps underpinned by somewhat paternalistic assumptions, Olmsted came to embrace a philosophy of urban public park design that gave great consideration to the needs of the underprivileged.

Olmsted's evolving notions in this regard were shaped by events that occurred prior to his career as a park designer. In the early 1850s, when Olmsted still thought that perhaps his professional future was as a tree farmer, he spent spent six months working on Fairmount—a 300-acre upstate New York farm owned by George Geddes.[29] Geddes, who became a mentor to Olmsted, was not the typical farmer of the day. He had studied the law and was an outspoken abolitionist who equated the cultivation of his land with the need to cultivate freedom and end slavery in southern states.[30] By Geddes's moral and religious beliefs, there was no justification for a system that deliberately stunted the development of America's blacks.

At the time he worked with Geddes at Fairmont, Olmsted likewise found the institution of slavery objectionable but believed that the South could be persuaded by economic reasons to phase out the practice over time.[31] He preferred not to frame the issue in Geddes's stark moral terms. Olmsted's gradualist approach to the slavery question began to give way in 1852, however, when he was commissioned by the *New York Daily-News* to write a series of travel articles (in the form of letters) on the cotton economy and culture of the American South.[32] In these newspapers letters, later published collectively in book form as *The Cotton Kingdom: Traveler's Observations on Cotton and Slavery in the American Slave States*, Olmsted adopted the nom-de-plume of Yeoman.[33]

Yeoman's early dispatches said little about the brutality of the slave plantation system but instead emphasized the lack of economic incentives for slaves to work hard or efficiently and contrasted this with the greater productivity of the northern workforce.[34] As Olmsted's travels through the American South continued, the tone and focus of Yeoman's letters shifted. He began to comment on the underdevelopment of white civic society where slavery predominated—the lack of libraries, colleges, and concert halls and literacy rates (among whites) that lagged far behind those in the North.[35] Yeoman's final letter published in the *New York Daily-News* shows little trace of the restraint of his early dispatches: "The North must demolish the bulwarks of this stronghold of evil by demonstrating that the negro is endowed with the natural capacities to make good use of the blessing of freedom; by letting the negro have a fair chance to prove his own cause, to prove himself a man, entitled to the inalienable rights of man. Let all who do not think Slavery right, or who do not desire to assist in perpetuating it, whether right or wrong, demand first of their own minds, and then of their neighbors, fair play for the negro."[36]

These notions of equity and fairness found later expression in Olmsted's approach to urban park planning. His focus was not so much on preserving native landscapes as it was on creating a naturalist setting that could provide a wilderness-type experience for those city citizens who lacked the means to experience more remote wilderness firsthand. At times, this meant Olmsted called for limiting recreational activities and people on park grounds to preserve wilderness-like elements—an insistence that in certain respects was arguably at odds with his pronouncements about how his parks would provide persons of lesser means with greater access to naturalist open space.[37] Therefore, notwithstanding his views about the social role of parks, on some occasions Olmsted's topographical concerns took precedence.

On the ground, the urban parks of Jensen and Olmsted may often look alike. The naturalist elements of Humboldt Park in Chicago and Central Park in New York City have much in common. But the philosophic and political soils from which the designs of these two parks grew are dissimilar, and this dissimilarity is relevant to current debates about urban rivers. In particular, this dissimilarity points to contrasting perceptions about who should control the urban river-planning process and about the interests of adjacent riverfront communities. These contrasting perceptions, in turn, factor heavily into the assessments of Moe, Wilkie, and Teaford concerning urban renewal's failings and into the statements of Villaraigosa and the other mayoral candidates at the September 2000 Occidental College forum.

The dissimilarities in the views of Jensen and Olmsted highlight the question of the beneficiaries of the greening American urban riverside lands. For whom is this greening being undertaken: For the minority and low-income residents who presently live and work near the riverfront where maritime and industry were formerly located? For the affluent white residents and workers who will move in following the gentrification that riverside parks will make possible? For the people who will come to revive themselves in a wilderness-like setting or for those who will engage in recreational activities like soccer, baseball, and frisbee? For the birds, fish, and mammals that will benefit from the habitat provided by restored wetlands and new woodlands? Our instinct may be to answer in the affirmative to all of these questions, but this instinct ultimately evades the reality that hard choices need to be made and that these choices may help some and hurt others.

A Poor Understanding

Proposals for urban river or creek ecological restoration projects in neighborhoods where primarily low-income and minority residents live are often dismissed as being unsupported by local residents. For these sites, the argument goes, the community's immediate interest is in creating new jobs and not in creating new parks in which residents can appreciate nature. If this claim is accepted at the political decision-making level, public resources for creekside or riverside restoration and parkland are redirected toward other communities (often with a different demographic composition), and land uses in low-income and minority neighborhoods continue to impair river and river-adjacent ecosystems.

This claim fits comfortably into Jensen's suggestion that most people who live in dense urban settings are incapable of appreciating native landscapes and natural resources. But this claim is disputed by many people presently involved in urban river-restoration efforts. As A. L. Riley notes in her 1998 book *Restoring Streams in Cities: A Guide for Planners, Policymakers and Citizens*, "An easy mistake to make is to assume that economically depressed, low-income neighborhoods, communities, and business districts have more pressing concerns. As a manager of the State of California's Stream Restoration Program, I saw some of the most innovative restoration projects occur in economically disadvantaged areas. These projects were not unique and isolated events but statistically significant. Such communities were using the cleanup and restoration of their inner-city creeks to improve their property values, attract businesses into the area, and strengthen older, centrally located business districts that had been on the decline."[38] Riley adds, "The greatest value of a restoration project may be the new sense of community identity or neighborhood pride created for the participants in the project."[39]

Riley's observations here are corroborated by findings in the 2004 report by the American Planning Association, *Ecological Riverfront Design: Restoring Rivers, Connecting Communities*. This report is primarily a technical guide to implementing ordinances, engineering approaches, and

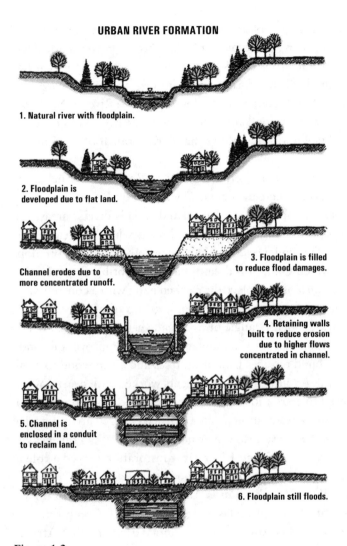

URBAN RIVER FORMATION

1. Natural river with floodplain.

2. Floodplain is developed due to flat land.

Channel erodes due to more concentrated runoff.

3. Floodplain is filled to reduce flood damages.

4. Retaining walls built to reduce erosion due to higher flows concentrated in channel.

5. Channel is enclosed in a conduit to reclaim land.

6. Floodplain still floods.

Figure 1.2
Diagram of urban river alternatives presented in the American Planning Association's *Ecological Riverfront Design*, 2004. Reprinted with permission of the American Planning Association; originally published in James Grant Mac Broom, *The River Book* (Hartford, Ct.: Natural Resources Center, 1998), 148.

design criteria for urban riverside lands. In addition to providing practical information on the zoning and ecology of urban rivers, however, the document also includes numerous case studies of low-income, minority urban districts taking the lead in efforts to reclaim degraded waterways and waterfront land, including communities along the Bronx River in New York City, the Swansea neighborhood along the South Platte River in Denver, and the Ravenswood neighborhood along the North Branch of the Chicago River.[40]

The restoration project on Berteau Street in Ravenswood helps put Riley's point in a more concrete context. The origins of this project, in one of Chicago's more economically challenged and racially diverse areas, was the collapse of a poorly engineered embankment where Berteau Street meets the water.[41] This collapse created an extremely steep drop-off along the river's edge, which made this area of bank susceptible to yet further erosion.[42] One option was to replace the earthen bank with a concrete wall. After consulting with ecologists, however, a group of neighbors chose instead to cut back the dense tangle of overhead vegetation, which would provide additional sunlight for groundplants and help improve soil conditions, and then to build terraces from dead trees and scrap wood to minimize bank erosion.[43] Additionally, the residents worked with nearby Water Elementary School to start a local environmental education program that used the Berteau Street restoration project to teach children about riparian plants, animals, and ecosystems in their neighborhood.[44] On this heavily urbanized stretch of the Chicago River, river restoration played a role in strengthening and stabilizing the Ravenswood community.

Another example of Riley's point is Augustus Hawkins Natural Park in the South-Central Los Angeles neighborhood of Compton. The park opened in December 2001 and was named after the first African American who was elected to Congress from California. In their initial discussions about converting an 8.5-hectare former municipal water-pipe disposal site to a park, many city planners assumed that the predominately lower-income African American and Latino residents who lived near the park would be interested primarily in basketball courts and soccer fields. Consultations with residents, however, showed that this assumption was wrong.

What the residents near the Hawkins Park wanted, aside from a space that was safe and free of gang activity, was a place where they could connect with nature.[45] The Pacific Ocean and the San Gabriel and Santa Monica Mountains were too far away for many local residents. In the tradition of Olmsted, Hawkins Park became a place not to preserve nature but to create nature for the benefit of inner-city residents who may lack the means to visit more remote wilderness. In designating Hawkins Park as a Great Public Space, the national nonprofit Project for Public Spaces explained: "As one drives down Compton Boulevard, trees become visible on the horizon. As you get closer to the park, the greenery stands out like a living beacon in a sea of concrete. There is ample seating along a path that circles the park. One of the main features of the park is it undulating topography, with hills and swales mimicking a native California setting. At the top of the hill, river rocks line a running stream, whose water is pumped by a windmill atop a hill, with water coursing down a small concrete spillway reminiscent of the L.A. River."[46]

In a 2002 article in *Landscape Architecture* magazine (published by the American Society of Landscape Architects), Trini Juarez, one of the landscape architects involved in the project, notes the prevalent yet misguided view that "underserved ethnic groups have no affinity for the outdoors."[47] The article's author comments, "Hawkins Park is a tangible rebuttal of many stereotypes about nature and the poor."[48]

All seven of the essays that follow in this book consider the questions of who makes decisions about our urban rivers, who pays to implement these decisions, and who ultimately benefits from or is burdened by these decisions. Therefore, to a certain extent, these essays pick up where Villaraigosa, Jensen, Olmsted, Riley, Ravenswood's Berteau Street neighbors, and Hawkins Park's creators left off.

The essays included were commissioned and selected based on four main criteria: geographic diversity so that the book was national rather than regional in scope; ongoing urban riverside land use disputes with uncertain outcomes to ensure the book's timeliness; varied institutional approaches, stakeholders, and problems to avoid redundancy; and authors with firsthand knowledge and involvement in the subject matter of their chapters. Without sacrificing scholarship, contributors have dirt beneath their nails, which allows their analyses of urban river-restoration efforts to reflect real-world experiences and not just library research.

In chapter 2, Robert Gottlieb and Andrea Misako Azuma of Occidental College's Urban Environmental Policy Institute introduce us to the strange

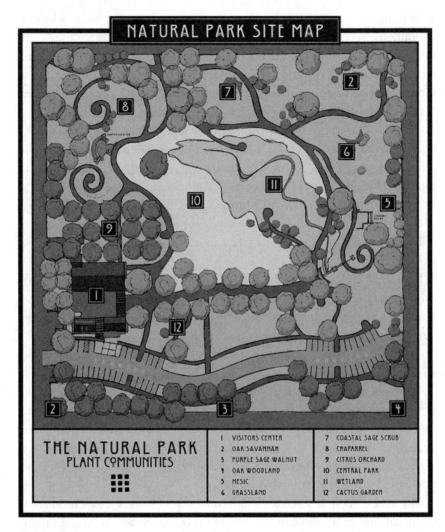

Figure 1.3
Informational poster for Hawkins Natural Park in Los Angeles, 2001. Courtesy of the City of Los Angeles and the Santa Monica Mountains Conservancy.

and evolving relationship between the City of Los Angeles and the Los Angeles River (formerly known as the Rio de Porciuncula). Once a naturally flowing waterway with a tendency to overflow its low banks and inundate large portions of the Los Angeles Basin, the river was paved and straightened by United States Army Corps of Engineers in the early 1950s, which transformed it into what has been described as a water freeway. In recent years, however, calls have increased to unentomb stretches of the Los Angeles River for the benefit both of riparian ecology and riverside communities. Gottlieb and Azuma describe the critical role that a local nonprofit group and a local academic research project have played in reenvisioning what the river is and might be.

Chapter 3, by Uwe Steven Brandes, former manager of the Anacostia Waterfront Initiative for the Office of Planning in the District of Columbia, discusses the lands along the Anacostia River. Unlike those along the Potomac River, the Anacosta's banks have been largely by-passed by the district's previous major planning efforts, such as the McMillan and L'Enfant plans. The waterfront areas and primarily African American neighborhoods along the Anacostia River have instead been the location of federal highway and urban-renewal projects that caused social disruptions that continue to this day. Within this setting, the Anacostia Waterfront Initiative has emerged as a vehicle for bringing attention and funding to this neglected section of the nation's capitol. Brandes provides an insider's account of the forces and processes that led to the Initiative's creation and analyzes its structure and operations to date.

Christopher Theriot and Kelly Tzoumis (in chapter 4) offer a case study of human interventions along with Chicago River, which naturally flows eastward through the City of Chicago toward Lake Michigan. The Chicago River lies just east of the westward-flowing waters of the Illinois River, which (unlike the Chicago River) is part of the Mississippi River watershed. In the early 1900s, Chicago city officials came up with an ingenious engineering solution to the problems of the city sewage overflows that were contaminating Lake Michigan. The Chicago River's flow into the lake was dammed, and a canal was built between the Chicago River and the Illinois River, thereby causing the river to reverse direction and sending the city's sewage overflows toward St. Louis along the Mississippi River. As Theriot (of Roosevelt University) and Tzoumis (of DePaul University) detail, this was the first in a long series of engineered interventions along the Chicago River, some of which are now being implemented to deal with the ecological consequences of the canal linking the Mississippi River and Lake Michigan watersheds. Although this chapter is focused more on instream impacts than most of the other essays in the book, its analysis of engineering-based solutions and effects on downstream communities picks up on common themes.

While the Los Angeles River may have been placed in a concrete straightjacket, this engineering solution seems mild compared to what happened in Salt Lake City. City Creek, a tributary to the Jordan River, had the misfortune of being located in an area slotted for downtown expansion. To facilitate this expansion, in the early 1900s City Creek was buried underground and for the past 100 years has been invisible. In part because of federal funding made available through brownfields programs and the new Urban Rivers Restoration Initiative, plans are now in the works to daylight this long-submerged waterway. In chapter 5, Ron Love of Salt Lake City's Public Works Department sheds light on the origins, agencies, and logistics involved in this daylighting effort.

Chapter 6, by attorney Richard Roos-Collins of the Natural Heritage Institute in San Francisco, looks at the Guadalupe River watershed. The Guadalupe and its tributaries make their way though Silicon Valley and the City of San Jose in California, eventually emptying into the south end of San Francisco Bay. Through his representation of a local resource conservation district, Roos-Collins was involved in litigation and an innovative settlement that seeks a long-term cooperative framework to address the problems of instream flow and water-quality impairment. The components of this settlement may serve as models for other urbanized areas facing similar river-related problems.

In chapter 7, Melissa Samet (senior counsel with the conservation group American Rivers) turns her attention to the federal agency that is at the center of much of this country's urban-river politics—the United States Army Corps of Engineers. The Army Corps planned and financed many of the large urban flood-control projects that channelized city rivers. Given the historical role played by the Army Corps in flood-control work, urban-river advocates are now looking to the agency to play a new role—that of restoring the rivers that it previously damaged. Samet gives us a sense of what has changed and not changed at the Army Corps on the urban-river policy front.

Finally, Mike Houck, in chapter 8, recounts the origins of the Coalition to Restore Urban Waterways (CRUW) and its Trashed Rivers conferences in the 1990s. As director of the Urban Greenspaces Institute and as the Portland Audubon Society's urban naturalist for more than twenty years, Houck is a veteran of efforts to restore bankside and instream wilderness along the heavily urbanized stretches of the Columbia and Willamette watersheds in the Portland, Oregon, metropolitan area. He was also part of a cadre of urban-river activists that coalesced a little over a decade ago to form a national movement. As Houck recounts, while CRUW and the Trashed Rivers conferences have disbanded, the legacy of these undertakings remains evident around the country.

Mike Houck's piece provides an apt theme on which to conclude. As Houck recalls, at the 1993 founding meeting for CRUW there was consensus that "although the urban stream movement would focus on urban waterways in general regardless of their location, this new organization would pay particular attention to streams and rivers in low-income, economically depressed areas. This nascent urban stream movement set its roots deeply and resolutely within an environmental justice matrix."[49] This movement can perhaps be viewed as a direct response to the urban-renewal experience, wherein the most immediately affected community residents were kept outside the land-use decision-making process.

The phrase *environmental justice*—the call for equal distribution of environmental benefits regardless of race or income—was not part of Olmsted's vocabulary when he undertook his city park projects in the 1800s. Yet it is not difficult to imagine Olmsted as a presenter on one of the Trashed Rivers conference panels, speaking against the elevated horseless-carriage promenades (highway overpasses) on our city riverfronts and insisting on a fairer allocation of urban riverside parkland resources. The terminology and technology would be new to Olmsted, and CRUW's activist rhetoric might strike his more paternalistic sensibilities as strange. The political landscape, however, would be all too familiar.

Notes

1. Beth Otto, Kathleen McCormick & Michael Leccese, *Ecological Riverfront Design: Restoring Rivers, Connecting Communities* 2–3, 97 (2004).

2. *See generally* Martin Anderson, *The Federal Bulldozer: A Critical Analysis of Urban Renewal 1949–1962* (1964).

3. Jon C. Teaford, *The Rough Road to Renaissance: Urban Revitalization in America 1940–1985* 17 (1990).

4. *Id.* at 18.

5. Richard Moe & Carter Wilkie, *Changing Places: Rebuilding Community in the Age of Sprawl* 50 (1997).

6. *Id.* at 57.

7. *Id.* at 68.

8. Teaford, *supra* note 3, at 245.

9. *Id.* at 232.

10. Transcription of September 2000 Mayoral Candidates Forum at Occidental College (on file with author), at 8–9.

11. *Id.* at 9.

12. *Id.* at 10.

13. Anne Whiston Spirn, *The Authority of Nature: Conflict and Confusion in Landscape Architecture, in Nature and Ideology: Natural Garden Design in the Twentieth Century* 249–62 (Joachim Wolschke-Bulmahn, ed. 1994).

14. Julia Sniderman Bachrach, *Jens Jensen: Friend of the Native Landscape*, Chicago Wilderness Magazine, Spring 2001, *available at* http://chicagowildernessmag.org/issues/spring2001/jensjensen.html (last visited Nov. 5, 2005).

15. *Id.*

16. *Id. See also* Charles E. Little's *Jens Jensen and the Soul of the Native Landscape, in* Jens Jensen, *Siftings* xiii (1939).

17. Robert E. Grese, *Jens Jensen: Maker of Natural Parks and Gardens* (1998).

18. Jensen, *supra* note 16, at 37–38.

19. Bachrach, *supra* note 14.

20. Jensen, *supra* note 16, at 28.

21. *Id.* at 30.

22. *Id.* at 96.

23. *Id.* at 26.

24. *Id.* at 60.

25. As discussed and quoted in Joachim Woldschke-Blumahn's introduction to *Nature and Ideology, supra* note 13, at 3.

26. Jensen, *supra* note 16, at xiii.

27. Witold Rbyczynsky, *A Clearing in the Distance: Frederick Law Olmsted and America in the Nineteenth Century* 177 (1999).

28. *Id.* at 173.

29. *Id.* at 64–66.

30. *Id.* at 66.

31. *Id.* at 107, 116.

32. *Id.* at 115–19; Frederick Law Olmsted, *The Cotton Kingdom: A Traveler's Observations on Cotton and Slavery in the American Slave States* 48, 171, 230, 262 (1953) (1861).

33. Rbyczynsky, *supra* note 27, at 119, 196.

34. *Id.* at 116.

35. *Id.* at 119.

36. *Id.* at 121.

37. *Id.* at 362–64; Frederick Law Olmsted, *Notes on the Plans of Franklin Park and Related Matters, The Papers of Frederick Law Olmsted, Supplementary Series,* Vol. 1, *Writings on Public Parks, Parkways and Park Systems* 478 (Charles E. Beveridge and Carolyn F. Hoffman eds., 1997).

38. Ann L. Riley, *Restoring Streams in Cities: A Guide for Planners, Policymakers and Citizens* 13 (1998).

39. *Id.* at 31.

40. Otto et al., *supra* note 1, at 34–35, 62–63, 125.

41. *Id.* at 125.

42. *Id.*

43. *Id.*

44. *Id.*

45. Kim Sorvig, *The Wilds of South Central: In South Central L.A., A New Park Tests Stereotypes about What Minority Groups Want from Parks,* Landscape Architecture Magazine, Apr. 2002, at 66–75.

46. *Available at* http://www.pps.org/gps/one@public_place_id=545 (last visited Aug. 19, 2005).

47. Sorvig, *supra* note 45.

48. *Id.*

49. Mike Houck, *infra* chapter 8.

2

Bankside Los Angeles

Robert Gottlieb and Andrea Misako Azuma

River Stories

"Landscapes tell stories," filmmaker Wim Wenders once declared, "and the Los Angeles River tells a story of violence and danger."[1] Wenders made these remarks during one of the forty sessions of the Re-Envisioning the L.A. River program, which was hosted by the Urban and Environmental Policy Institute at Occidental College in 1999 and 2000. Wenders's comments were made during a panel discussion on how Hollywood films presented the Los Angeles River as a backdrop for the stories those films told.[2] The year-long program was organized to explore how the discourse about the river could be changed to reorient the policy framework regarding the management and future design of this heavily engineered and reconstructed urban waterway and the related land-use considerations for the areas bordering it.[3]

The landscapes that Wenders referred to were from a film montage entitled *River Madness* that was edited to include various L.A. River scenes from such Hollywood movies as *Grease*, *Terminator 2*, *Repo Man*, and *Them*.[4] *Them*, a classic 1950s science fiction film that depicted giant irradiated ants crawling out of the storm drains that fed into the L.A. River, had been filmed at a point in time when the "declaration of war on the L.A. River," as one U.S. Army Corps of Engineers official characterized it, was reaching a conclusion.[5] By the 1950s, a new landscape had been constructed—a channelized river that served as a passageway for unwanted floodwaters and scattered debris.[6] This new river, or flood-channel "freeway," told a story of a dangerous, polluted, and fragmented Los Angeles, a place barren of the softer, more inclusive landscapes of green and open space often associated with images of nonurban rivers.[7]

During the past decade, the L.A. River has become a subject of intense reexamination, a major topic of policy debate, and a new kind of environmental icon. It has increasingly come to symbolize the quest to transform the built *urban* environment from a place seen as representing violence and hostility for communities and for nature to a place of rebirth and opportunity.[8] Reenvisioning the Los Angeles River as a place of community and ecological revitalization (rather than an exclusive and dangerous flood channel fenced off from surrounding communities) provides a powerful message of renewal for urban rivers and the quality of urban life.[9] It also provides lessons in how institutional and policy changes can be influenced by the way an issue is framed—whether in relation to its historical context, its environmental and economic aspects, or its relationship to the broader discussion of land use at the local and regional levels.[10]

This essay explores some of the influences on that process of reexamination. It includes a discussion of the roles of a community-oriented academic entity (the Urban and Environmental Policy Institute) and a nonprofit organization (the Friends of the Los Angeles River) whose long-standing mission has been to enable policymakers and residents alike to rediscover this urban river. We also reflect on how the changing discourse around the river helped advocates mobilize support and influence policies in support of community and ecological revitalization.

"The River We Built": A River Transformed

In 1985, *Los Angeles Times* writer Dick Roraback documented his exploration of the Los Angeles River.[11] The intent of his twenty-part series, published over several months, was to travel up the river from its mouth to its source.[12] This reverse process of discovery was designed in part to answer the questions "What is the L.A. River?" and "Where does it begin?"[13] The tone of the article was partly comic and partly ironic, and it took the form of a third-person narrative, with Roraback characterizing himself as "the Explorer."[14] Roraback began his series at the mouth of the river in Long Beach.[15] With its soft bottom area not fully channelized, Roraback suggested that a river might in fact exist:[16] "For a last time, the Explorer looks south, at the real Los Angeles River. A heron-like bird, easily four feet tall, stands motionless in the stream, graceful and haughty," Roraback wrote, as he headed upstream.[17] But as the soft bed changed to flat, concrete sides, and Roraback crossed the old industrial section southeast of downtown Los Angeles that consisted of some of the densest communities in the region, the *L.A. Times* writer described a "desolate vista, a wasteland . . . just a threadbare coat of unspeakable slime."[18] Drawing on river analogies and poking fun at every opportunity about the degraded river and its nearly nonexistent flow, Roraback asked, "Is 'Old Man River' in drag?' Is the 'Beautiful Blue Danube' in a mudpack?"[19] When he finally reached what he assumed was the river's source, he complained that he never really did find what could be considered a river, since, "It hasn't any

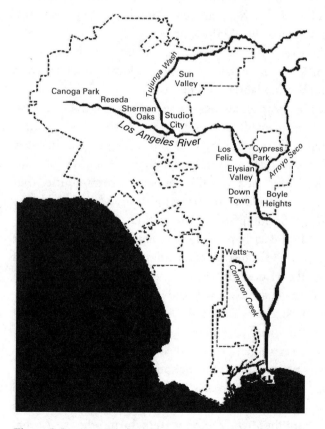

Figure 2.1
Map of the Los Angeles River watershed, October 2002. Reprinted with permission of the Los Angeles City Council Ad Hoc Committee on the Los Angeles River.

whitecaps. It hasn't any fish. . . . Just to see one ripple would be my fondest wish."[20] Instead, the Explorer moaned, the L.A. River "just hauls its load of sad debris from the sewage pipes to the mighty sea." "Ooze on, L.A. River, ooze on," Roraback concluded his ironic homage to this forgotten and bleak part of the Los Angeles landscape.[21]

When Dick Roraback undertook his journey, the Los Angeles River, as a free-standing river, had become more memory than reality.[22] The L.A. River, in fact, has experienced multiple lives over different eras as a pueblo became a boom town and eventually a continually expanding urban land mass.[23] "Making the river a critical part of the landscape made sense in the early days of the little village's history," historian William Deverell wrote of the pueblo period during the eighteenth and nineteenth centuries:[24] "Local knowledge, based on lived experience in the Los Angeles basin, incorporated the river into the rhythms of everyday life. Los Angeles needed no more water than the river could provide, and it was an especially prominent landscape feature along with other local markers such as the Pacific Ocean or the San Gabriel Mountains."[25]

Today, L.A. River restoration advocates seek to invoke the historical image of a free-flowing river filled with willows, cottonwoods, watercress, and duckweed that was "a very lush and pleasing spot, in every respect," as its first Spanish chronicler, Father Juan Crespi, wrote in his diaries back in 1769.[26] But the river also served the communities that grew up around it. It was used as an irrigation source for agricultural land (including during flood episodes) throughout much of the nineteenth century,[27] and in the early part of the twentieth century, the river received discharges from the industrial plants that lined its edge as part of the East Side industrial corridor.[28] In 1930, it became the centerpiece of a Chamber of Commerce–commissioned study by Harlan Bartholomew and Frederick Law Olmsted Jr.[29] This study—which presented a vision of greenbelts, parkways, and new park lands—was as forgotten as the L.A. River itself until it was revived and republished by Hise and Deverell seventy years later.[30] Most significantly, throughout much of the early twentieth century, the L.A. River came to be seen as a barrier for existing and future residential and industrial development along its path, due to its propensity to carry rapidly flowing flood waters during the occasional but fierce storms that periodically occurred.[31] Two major storms in 1934 and 1938 helped facilitate the entry of federal dollars (part of a broader

New Deal job-creation strategy associated with public-works projects) to initiate a wide range of construction projects to effectively (and finally) manage the river to prevent future flooding.[32]

From 1938 (when the Army Corps began to straighten the river by constructing a channel along much of its fifty-one miles) through the 1980s (when Dick Roraback set out to find his lost river), the L.A. River became transformed into a flood-control throughway to carry water to the ocean and prevent flooding.[33] Similar to the flood-control projects that also sought to reconfigure urban streams and rivers around the country during this same period, the now channelized L.A. River essentially redefined the urban landscape along a north-south axis.[34] Areas surrounding the river were now fenced off and became a forbidden territory that effectively belonged to the engineering agencies.[35] For the flood-control managers, this was now "the River we built," as one Army Corps engineer described it.[36]

"Bring the River Back to Life": Friends of the L.A. River

At the same time that Dick Roraback was publishing his twenty-part series, a poet and performance artist named Lewis MacAdams sought to make a very different kind of discovery about the Los Angeles River through his poetry and art.[37] In an act that was part theater and partly an action designed to spur organizing, MacAdams, with three artist colleagues, cut through the fence at a location on the river just north of downtown Los Angeles close to where one of the soft bottom areas gave way to the concrete channel.[38] Entering the channel, he proclaimed that the river still lived below the concrete.[39] "We asked the river if we could speak for it in the human realm. We didn't hear it say no," MacAdams would later comment on his act on a number of occasions.[40]

From this event, MacAdams formed a new organization, Friends of the L.A. River (FoLAR), whose initial goal was to focus on language and symbols by insisting that the L.A. River was indeed a river.[41] MacAdams's activist roots were more bound up with his identity as poet and affinity for imaginative 1960s-style protest than any specific environmental or river advocacy lineage. He tended to attract like-minded artists, planners, architects, designers, and neighborhood activists in this quest to "bring the River back to life," as he wrote in a letter to the editor of the *L.A. Times* in

Figure 2.2
The Los Angeles River near the Glendale Narrows following a heavy storm, January 2005. Photograph by Paul Stanton Kibel.

response to Roraback's series.[42] Roraback, who put a FoLAR bumper sticker on his car while undertaking his river journey, nevertheless identified as hopeless the task of the group he called "Sons of the Ditch."[43]

For the engineer-managers in the L.A. County Department of Public Works and the U.S. Army Corps, the emergence of FoLAR was, at first, more of a nuisance than a significant challenge to the roles they had assumed as flood-control managers.[44] Three sets of events, however, escalated FoLAR's theatrical protests into a war of words and symbols that ultimately influenced the policy and institutional framework of river management and related land-use issues.

The first event involved the flow of the river itself. In 1984, the Los Angeles Department of Water and Power began operating its Donald A. Tillman Water Reclamation Plant north of the Sepulveda Dam near the source of the river and one of its three soft bottom areas. This tertiary sewage-treatment plant discharged directly into the river.[45] Along with the releases of two other smaller treatment plants south of Tillman, these discharges provided a year-round flow of water for the River.[46] As a consequence, it increased the vegetation growth and habitat along the soft bottom areas and reinforced the FoLAR argument that, at least in these stretches of the river, visually and functionally (in relation to a renewed ecosystem), the L.A. River had become once again a free-flowing river.[47] Even with the negative symbolism of treated sewage as its water source (a source nevertheless cleaner than the runoff flow, given the tertiary treatment process involved and the large pollutant loads in the runoff), this new river flow reinforced FoLAR's appeal about a living river.[48] "Come down to the River," became a constant refrain in talks by MacAdams and his FoLAR allies, an action they considered essential to legitimating their argument about the river.[49]

The second event involved a concept put forth by then State Assembly member Richard Katz. In 1989, Katz proposed that the river, much of it channelized and lacking any human contact, could serve as a "bargain freeway" for trucks and automobile traffic.[50] The river-freeway concept was further explored through a $100,000 L.A. County transportation study that concluded that such a river freeway could result in a 20 percent reduction in congestion for two nearby freeways.[51] Katz, who was a major advocate of water transfers and sought to appeal to environmental groups in his subsequent run for mayor and the state senate, also spoke of greenbelts, bikeways, and adjacent parks—ideas that had been promoted by FoLAR and its allies.[52] Katz's argument about a river freeway, never seriously pursued and eventually ridiculed by the media, nevertheless created a new kind of focus on the river that FoLAR and its allies were able to exploit.[53] "Why not re-envision the L.A. River as an actual river?" the activists argued in documents and materials they generated, in events they hosted (such as river clean-up days and kayak rides along the soft bottom areas of the river), and in the increasing number of press interviews and articles that identified the FoLAR vision of a living river.[54]

The third event involved the protracted battle over the Army Corps's proposal to raise the channel walls in the downstream segment of the river prior to its entering the Long Beach Harbor area (the same segment of the river that Dick Roraback had characterized as a "wasteland").[55] In 1987, the Army Corps produced an update for its L.A. River master plan for the Los Angeles County Drainage Area (LACDA) that included the warning that disastrous flooding could return to Los Angeles County.[56] The Corps proposed a series of measures to address a number of problems that had emerged postchannelization.[57] These included increased residential development along the river's edge, debris-flow concerns, and emerging fears about flood-damage insurance for homeowners due to FEMA's declaration of certain areas bordering the river as "flood hazard zones."[58] Asserting that neighborhoods bordering the L.A. River, particularly those downstream in the "wasteland" areas, required protection from a 100-year flood, the Corps plan included widening the channel, modifying bridges, and, most controversially, constructing new parapet walls from two feet to as much as eight feet higher than their current height.[59]

Almost immediately, the Army Corps's LACDA proposal became a flash point for FoLAR and other river advocates who sought to challenge the Army Corps's approach while identifying their own L.A. River flood-management and restoration plan as a counterpoint to the raising of the walls.[60] The credibility of the river advocates was also subsequently enhanced by their participation in a Los Angeles River Task Force, established in 1990 by Mayor Tom Bradley to "articulate a vision for the River."[61] Through the early 1990s, the LACDA fight became the centerpiece of the debates over the future of the river.[62] On the one hand, the engineering agencies (the Army Corps and the L.A. County Department of Public Works) argued that the LACDA proposal was simply an extension of their mission as flood-control managers and that to call the river a "river" was a misnomer.[63] The river, they declared, served just two purposes—to keep flood waters from destroying property and lives ("a killer [that was now] encased in a concrete straight jacket," as one water-agency publication put it) and to manage the discharges from the sewage-treatment plants.[64] FoLAR countered with a series of alternative management strategies that included the twin concepts of "restoration" (tearing up the concrete where feasible) and "flood management."[65]

In one memorable encounter, described by Blake Gumprecht in his history of the L.A. River, Jim Noyes (the chief deputy director of the L.A. County Public Works Department) got into a sharp exchange with FoLAR's MacAdams over what term to use when describing the river.[66] Each time Noyes used the term "flood-control channel" as part of a presentation he was making, MacAdams would interrupt to declare "you mean 'river.'"[67] This happened again and again, with Noyes insisting on using the term "flood-control channel" and MacAdams interrupting each time to assert "river!" MacAdams later recalled the incident as turning "really ugly," with Noyes becoming more and more furious.[68] "I saw him a couple of days later," MacAdams told Gumprecht, "and he wouldn't even speak to me."[69]

Despite FoLAR's newly established visibility and the increased interest by policymakers and the media in a "renewed" Los Angeles River, the County Board of Supervisors in 1995 voted four to one to allow the LACDA plan to proceed.[70] Given the changing discourse around the river and the focus on the limits of traditional flood-control strategies as well as concerns about community blight, the LACDA plan was modified to the extent that in some areas levees were raised and walls were not built.[71] In addition, adjacent bike paths were maintained, and more plantings and vegetation were added to counter the "urban blight" and "wasteland" characterization of the engineering approach.[72] And although FoLAR had lost a battle, it continued to make inroads in how the river was to be defined and how it might ultimately be managed and renewed.[73] "I always saw [the LACDA fight] as a symbolic issue," MacAdams recalled, arguing that the war of words was in fact "a battle over the definition of the river and what the river is going to be."[74]

"A Very Pretty Duck": The Reenvisioning Program

In the fall of 1998, one year after the Los Angeles County Board of Supervisors voted on LACDA, Lewis MacAdams came to Occidental College to speak to an environment and society class about the history of the L.A. River.[75] In the course of a discussion with students, MacAdams commented that a key dimension of FoLAR's approach was changing the image of the river—an image, MacAdams speculated, that might have been shaped in part by its portrayal in Hollywood films.[76] MacAdams argued that research was needed into how such images and language about the river were formed, how the river evolved historically, how it could be reengineered differently, and how planning could reconfigure not just the river but its surrounding communities.[77] He challenged his audience: "Today, the research regarding the River serves as barrier rather than opportunity for renewal; can you help make that renewal possible?"[78]

MacAdams was also interested in a possible partnership with Occidental's Urban and Environmental Policy Institute (UEPI).[79] It seemed a good fit.

UEPI's predecessor (an interdisciplinary Environmental Center first organized in 1991 at UCLA that linked the departments of chemical engineering, public health, and urban planning) drew on the technical and research capacity of different disciplines and was established as an "action research" program to help develop new public policies and establish linkages with key stakeholders, including community-based organizations.[80]

In 1997, Robert Gottlieb (director of the UCLA Environmental Center) and three project managers from the UCLA center moved to Occidental College and brought with them their projects and programs. Occidental is a small, highly diverse liberal arts college located in the heart of Los Angeles, not far from where the L.A. River begins to enter downtown Los Angeles. While the work of the UCLA Environmental Center had been innovative and productive with a strong community emphasis, it often found itself operating at the margins of or outside the academic program. At Occidental, however, the Urban and Environmental Policy Institute (UEPI) became a centerpiece of Occidental's own commitment to community engagement and "learning by doing."[81] Thus, while UEPI was situated within an academic program (urban and environmental policy), it defined itself as being oriented toward social change and as providing a place where faculty, students, organizers, community partners, researchers, and policy analysts could collaborate. Its mission—"to help create a more just, livable, and democratic region"—became the backdrop for bringing together community groups and researchers.[82] It therefore saw itself as a cross between an academic center with strong community ties and a community-based organization with a strong research and policy-development capacity.

By 1999, UEPI and FoLAR decided to pull together a year-long program of events, activities, forums, and research under the heading Re-Envisioning the L.A. River. More than forty programs were scheduled—a forum on the history of the river, a dialogue between the engineer managers of the U.S. Army Corps and L.A. County Public Works and alternative and environmentally oriented "flood-management" engineers and advocates, a presentation about possibilities for river renewal by the two leading environmental officials in California (Mary Nichols, secretary of the California Resources Agency, and Felicia Marcus, administrator of Region IX of the U.S. Environmental Protection Agency), and a meeting hosted by the mayor of the city of South Gate to discuss river-renewal and community issues south of downtown in what were called the Gateway communities

(Dick Roraback's wasteland areas).[83] There was also a poetry reading organized in conjunction with the Getty Research Institute (seven leading L.A. poets commissioned to write about the river), an art installation along the concrete walls in the area where Lewis MacAdams had proclaimed fifteen years earlier that a river still lived beneath that concrete, a bike ride along the river cohosted by the L.A. County Bike Coalition, and the Hollywood Looks at the River forum where the *River Madness* video montage was screened and Wim Wenders spoke of the impact of the river as a landscape of violence and danger.[84]

The research that was generated by the Re-Envisioning the L.A. River series included an historical reconstruction of the ecology of the Arroyo Seco (a twenty-two-mile stream and subwatershed that feeds into the L.A. River watershed), which was designed to place in context the limits and opportunities of stream-restoration strategies based on that historical information.[85] It also included client-based research where a group of UCLA urban-planning graduate students worked in conjunction with UEPI (as the client) to provide a community and planning profile of a parcel along the river known as the "Cornfield."[86] This hotly contested area just north of downtown Los Angeles was slated for warehouse development, much to the dismay of river activists and area residents, and the UCLA report outlined strategies and alternative scenarios for future development.[87] The battle over the fate of the Cornfield, another key topic of the Re-Envisioning the L.A. River program, was emerging in 1999 and 2000 as the first major debate over the fate of the river and its surrounding areas since the protracted battle over LACDA and the raising of the walls that had occurred during the early and mid-1990s.[88]

The opening session of the Re-Envisioning the L.A. River program on October 1, 1999, included talks by environmental officials Mary Nichols and Felicia Marcus. Nichols spoke of the role that could be played by forums like the Re-Envisioning series in focusing the attention of policymakers on open space and river revitalization as both community issues and environmental concerns.[89] Marcus talked of the importance of combining vision with the ability to act in a practical and sometimes incremental manner, praising the Re-Envisioning series for attempting to develop those kinds of links.[90] She further spoke of the need to establish new management paradigms while recognizing and gently pursuing a shift in the traditional agendas of the engineering, water-industry, and flood-control

Figure 2.3
Freight trains at the Cornfield site, circa 1930. North Broadway Bridge over the Los Angeles River in the distance. Courtesy of the California Railroad Commission and California State Archives.

actors who had managed the river for more than six decades.[91] While the river might not be "the [trumpeter] swan in L.A.'s future," Marcus said while summing up the event's message, "it could be a very, very pretty duck," citing *Los Angeles Weekly* writer Jennifer Price's compelling metaphor of a reenvisioned river in an article about the series that had appeared shortly before the Nichols and Marcus talk.[92]

New Battlegrounds: From Discourse to Action

Later that evening, after the initial session with Mary Nichols and Felicia Marcus, a group of FoLAR activists sat down with UEPI staff and Nichols and Marcus at an Eagle Rock neighborhood restaurant to talk about the future of the river. The FoLAR and UEPI participants told Nichols and Marcus about an emerging conflict regarding the Cornfield site. A year earlier, FoLAR had hosted the River through Downtown, a conference involving urban architects and designers, participants from the Chinatown community, and various river advocates.[93] FoLAR had been excited about the community involvement that had come out of the process, facilitated in part by a Chinatown activist named Chi Mui, who then was State Senator Richard Polanco's Chinatown field deputy. Mui (who had his own activist roots but was not at the time a river advocate and was not focused on environmental issues) nevertheless came to see the development of the Cornfield, a forty-acre site then owned by Union Pacific Railroad near the Chinatown and Latino Lincoln Heights neighborhoods north of downtown, as a major opportunity to address a wide range of community needs.[94] Through discussions with community members and through the design and envisioning process from the River through Downtown conference and its aftermath, FoLAR, Chi Mui, and a number of community participants came up with a plan for schools, housing, bike paths, recreational facilities, a park, and a more extensive riverfront development that could be ultimately tied to the broader vision of river renewal.[95]

But both FoLAR and the Chinatown activists soon discovered that the process they had launched would be undercut due to a very different kind of development that had been proposed by a large developer, Majestic Realty, for thirty-two of the acres of the Cornfield site. Majestic was

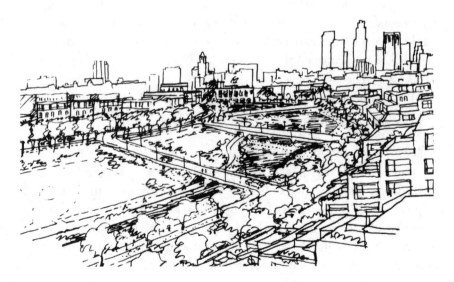

Figure 2.4
Sketch of the Los Angeles River Corridor presented at the River through Downtown conference, 1998. Reprinted with permission of John R. Dale.

considered one of the most politically powerful developers in the region. It had worked out a deal with Union Pacific (whose owner also had an interest in the development) to take over the land and effectively turn the site into a new warehouse and light industrial district.[96] Majestic also had ties and the strong support of then Los Angeles Mayor Richard Riordan.[97] Its development plan was on a fast track through the city and the federal government's review process and was expected to obtain subsidies to sweeten the deal.[98]

Neither Nichols nor Marcus had heard about Majestic's warehouse development, which had been pursued largely under the public radar.[99] Despite the increasing ability of river advocates to stimulate interest in river renewal, the FoLAR community plan seemed dead in its tracks, given the forces pushing the Majestic Realty plan. When Nichols and Marcus heard about this possible new battleground, they warned the river advocates that by taking on some formidable powers, they needed, in Nichols's words, to "slay the King, if they were going to win the battle."[100]

The Cornfield battle, in fact, identified a new stage in the advocacy around river renewal. Enlisting the support of a wide range of community and environmental organizations, evoking historical and cultural arguments about the significance of the site, and employing a range of legal and lobbying strategies to block Majestic's fast track to development, river advocates displayed a new level of sophistication and capacity to act.[101] The conflict, it was argued, was between environmental and community needs on one side and on the other side a development plan that would bring more pollution and poor land-use planning to that community.[102]

The Re-Envisioning the L.A. River program also played a role in the unfolding Cornfield dispute. The UCLA/UEPI research report that evaluated the competing visions about the past and future of the Cornfield helped provide documentation about the needs of the surrounding communities and the negative impacts associated with the Majestic Realty proposal.[103] In September 2000, the final event in the Re-Envisioning series, a mayoral candidates' debate about the L.A. River and the urban environment, was an animated discussion about the Majestic Realty project, alternative scenarios about the site, and general river renewal issues.[104] Each of the candidates present either declared opposition to the Majestic project or sought to slow down the fast-track approach, either by having the federal government's Housing and Urban Development (HUD) agency require an environmental review or by requiring a mediated dialogue between FoLAR and Majestic. The mayoral candidates' debate in turn suggested that the political climate around the project had significantly changed. "It is hard to adjust to the fact that the L.A. River has become a kind of mom and apple pie issue," MacAdams commented to Gottlieb immediately after the debate.[105]

In the several months after the September 2000 debate, negotiations took place between the developer and state of California officials over the price and conditions of a sale of the Cornfield property to the state.[106] Ultimately, a deal was reached, significantly benefiting the developer in terms of the final price but also making available undeveloped property along the edge of the river to be transformed into a state park.[107] Though the plans that were subsequently proposed (with a final decision still pending) did not fully coincide with the vision of the river advocates, in just four years (from the vote on the LACDA proposal to the resolution of the Cornfield battle), the power of the river advocates had grown to the point where they had crossed the line from discourse to action.[108]

Continuing Challenges: South of Downtown

In developing the Re-Envisioning the L.A. River program, the river advocates discovered that the potential north-south divide (which had been a major factor in the LACDA fight) was a difficulty. Although some of the communities north of downtown, including those affected by the Cornfield issue, were diverse in terms of demographics and included a number of low-income neighborhoods, the river advocates were at times put on the political defensive as middle-class advocates seeking to increase environmental amenities along the river.[109] These amenities were made possible by the images of renewal associated with the soft bottom areas. The call to "come down to the River" provided an important organizing tool in those communities, in part because it was possible to actually enter the river's edge and imagine a more robust, free-flowing river. Strategies for river restoration also seemed more viable in these areas.

One program highlighted during the Re-Envisioning program included a panel that evaluated a program that had established a recreated stream along the Arroyo Seco subwatershed.[110] This type of "reinvented Nature" (to use environmental historian William Cronon's suggestive phrase) had been made possible two years earlier by mitigation funds from a landfill developer.[111] The panel's discussion of the recreated Arroyo Seco diverted stream pointed to the significant challenges but very real opportunities tied to Felicia Marcus's call for practical, incremental steps to establish new river "management paradigms."[112] These opportunities, however, seemed limited if not nonexistent in the southern part of the north-south divide, particularly in the areas where the walls had been raised.[113] To reenvision the channelized river in this more desolate landscape seemed a nearly impossible task.

However, a number of events and political shifts that were unfolding at the time of the Re-Envisioning program suggested new openings that had not previously been available. In March 2000 and again in 2002, two bonds for park land acquisition and water-quality improvement projects provided for the first time significant funding for urban park and recreational development.[114] Both bonds passed with large majorities of Latino and African American voters, whose support exceeded that of white voters as well.[115] Funds from the March 2000 bond measure were subsequently used to acquire the Cornfield property. In late 1999, legislation was also passed that established the San Gabriel and Lower Los Angeles Rivers and Mountains Conservancy.[116] This new entity provided potential resources as well as research and planning opportunities for the segment of the river south and east of downtown Los Angeles as well as for the San Gabriel River and Rio Hondo subwatershed that joined the L.A. River in the city of South Gate.[117] One of the Re-Envisioning events, hosted by State Senator Hilda Solis, who was preparing for a successful run for Congress, had focused on the San Gabriel River.[118] Similar to other Re-Envisioning the L.A. River discussions, this event addressed the parallel issues of river renewal, community needs for open space and recreation, and broader land-use strategies that extended beyond the river's edge.[119] These events and other indications of an emerging community-based environmentalism that identified strong Latino, African American, and Asian American interest in open space and recreational development in urban areas extended the potential base for a new approach to river renewal.[120]

In recognition of these possibilities, FoLAR worked closely with a team from Harvard University's department of landscape architecture to extend the vision of river renewal to the channelized segments through downtown Los Angeles and southward to the L.A. city limits at Vernon. The charge of the Harvard team was to create a "provocative vision of the Los Angeles River that challenges the viewer to imagine a dreamscape of beaches, new ecologies, and connections across the city." In doing so, the Harvard design team considered how river managers might be able to "bring back the habitat, clean the water, and make it a natural amenity, while maintaining flood protection," with the goal of transforming the river "from today's poor joke into the centerpiece of a great city."[121]

The Re-Envisioning the L.A. River series also sought to focus on the south of downtown segment of the river through a panel discussion and community forum cohosted by the City of South Gate. In advance of this program, an Occidental College student, who grew up and went to school in South Gate, did presentations and outreach activities, through UEPI, with high school students in the area.[122] She found an interested and receptive audience that was largely unaware of the existence of the river at the edge of their city but interested in exploring a visioning process that could enhance their neighborhood.[123] Nevertheless, the high school students informed the Occidental organizer they had no interest in the community forum, since it was "a city thing."[124] The community forum included the mayor of South Gate (himself a board member of the new Conservancy), an environmental consultant who provided information on the problems and extent of contamination at the different brownfield sites along the river's edge in the south of downtown region, and a sprinkling of public officials. Poorly attended (few South Gate residents attended despite the efforts by UEPI and the city to generate an audience), the event suffered from a disconnect between the reality of a concrete river and a visioning process that up to then had little connection to the needs of those communities south of downtown.[125]

NGO and Academic Influences on the Terms of the Debate

From 1985, when Dick Roraback took his expedition upriver and Lewis MacAdams undertook his performance art by entering the river channel, to the September 2000 mayoral debate as the concluding event in the Re-Envisioning the L.A. River series, the key to bringing about change in the

Los Angeles River was the need to change the terms of the debate about how people viewed this highly engineered urban river. The debate and the actions that ensued constituted a "discourse battle" over how language was used that in turn framed an issue, identified resources, and established new practices and policies. In that context, FoLAR, a nongovernmental organization well equipped to influence the terms of that debate, became the leading actor in reorienting policy and institutional approaches. Similarly, UEPI, an academic entity that also functions in part as a community actor, could play a significant role in combining its research and educational functions, its cross-disciplinary approach that placed river issues in multiple contexts (historical, poetic and artistic, engineering, and political), and its community outreach and policy development functions. By 2000 and 2001, with the successful outcome in the Cornfield fight, it was clear that the discourse battle had not only been joined but that the terms of the debate had changed.

One example of this shift in discourse was reflected in D. J. Waldie's evocative commentaries about the L.A. River. Waldie, a novelist and city official for one of the cities south of downtown that was most vulnerable to floodwaters, frequently wrote about the river in the opinion pages of the *Los Angeles Times*. In an *L.A. Times* commentary written shortly before the opening event of the Re-Envisioning the L.A. River series, Waldie described the San Gabriel and Los Angeles Rivers as "problematic." [126] "The gated and trespass-forbidden river channels seem superfluous, the ultimate 'no place' in notoriously placeless L.A.," Waldie wrote. [127] Reflecting that shift in discourse around the river, Waldie's position also evolved, reflected in another *L.A. Times* commentary soon after the Re-Envisioning the L.A. River series had concluded. "As we begin to encounter the river as a place, not an abstraction, we encounter each other," Waldie wrote. [128] "The riverbank is not the perfect place for this meeting, but it's the only place we have that extends the length of metropolitan Los Angeles and along nearly all the borders of our social divides. Think of the river we're making as the anti-freeway—not dispersing L.A. but pulling it together." [129]

A few years later, Waldie, now a champion of river renewal, noted in a 2002 *New York Times* commentary that "recovering parks from industrial brownfields" wouldn't "restore a lost Eden," given that the greening of the Los Angeles River was "a sobering demonstration of the limits of environmental restoration in an urban landscape."[130] But the major accomplishment of actors like FoLAR and UEPI had been to help start the process of enabling the engineers, policymakers, and community residents to change agendas and establish those new management paradigms, while seeking to reverse what had also seemed to be an inexorable outcome of closeting the river and dividing a city and a region. "It has been the nature of Angelenos to be heedless about their landscape," Waldie concluded in his *New York Times* commentary:[131] "That's changing, because it must, as we finally gather at the river."[132]

Notes

1. *Hollywood Looks at the River*, videotape (CBS Studio City 2000) (on file with authors). During the discussion, Wenders also commented that "landscapes ask for their own stories to be told. The L.A. River, as it now exists as a cemented river, has a story of aggression to tell."

2. Robert Gottlieb & Andrea Azuma, *Re-Envisioning the Los Angeles River: A Program of Community and Ecological Revitalization*, Council for the Humanities (2001), *available at* http://organizations.oxy.edu/lariver/publications/Re-envisioning%20the%20LA%20River%20Community%20and%20Eco%20Revitalization.pdf (last visited Aug. 15, 2001).

3. *Id.*

4. *Grease* (Paramount Pictures 1978); *Terminator* 2 (Carolco Pictures, Inc. 1991); *Repo Man* (Edge City 1984); *Them* (Warner Brothers 1954).

5. Andrew Boone, *River Rebuilt to Control Floods*, Scientific American, Nov. 1939, at 265. *See also Flood Control Program for Los Angeles*, Western Construction News, Nov. 1939, at 148. The *River Madness* film was produced by Dana Plays for the panel discussion.

6. Jared Orsi, *Hazardous Metropolis: Flooding and Urban Ecology in Los Angeles* 101–02 (2004).

7. Gottlieb & Azuma, *supra* note 2, at 2.

8. *Id.*

9. *Id.*

10. *Id.*

11. Dick Roraback, *Up a Lazy River, Seeking the Source Your Explorer Follows in Footsteps of Gaspar de Portola*, L.A. Times, Oct. 20, 1985, at 1. The series ran intermittently between October 20, 1985, and January 30, 1986.

12. *Id.*

13. *Id.*

14. *Id.*

15. *Id.*

16. *Id.*

17. Dick Roraback, *The L.A. River Practices Own Trickle-Down Theory Series: In Search of the L.A. River*, L.A. Times, Oct. 27, 1985, at 1. *See also supra* note 11.

18. Dick Roraback, *Bridging the Gap on the L.A. River with a Song in His Heart and a Yolk on His Shoe Series: In Search of the L.A. River*, L.A. Times, Nov. 7, 1985, at 37. *See also supra* notes 11, 17.

19. *Id.*

20. Dick Roraback, *From Basin Camp, the Final Assault Series: In Search of the L.A. River*, L.A. Times, Jan. 30, 1986, at 1. *See also supra* notes 11, 17, 18.

21. *Id.*

22. Orsi, *supra* note 6, at 102.

23. William Deverell, *Whitewashed Adobe: The Rise of Los Angeles and the Remaking of Its Mexican Past* 99 (2004).

24. *Id.*

25. *Id.*

26. Blake Gumprecht, *The Los Angeles River: Its Life, Death, and Possible Rebirth* 38 (1999).

27. *Id.* at 142.

28. Greg Hise, *Metropolis in the Making: Los Angeles in the 1920s, in Industry and Imaginative Geographies* 40 (Tom Sitton & William Deverell eds., 2001).

29. Greg Hise & William Deverell, *Eden by Design: The 1930 Olmsted-Bartholomew Plan for the Los Angeles Region* 52 (2000).

30. *Id.* at 52.

31. Orsi, *supra* note 6, at 88–92.

32. Robert Gottlieb, *Environmentalism Unbound: Exploring New Pathways for Change* 18 (2001).

33. Orsi, *supra* note 6, at 102.

34. Ann Riley, *Restoring Streams in Cities: A Guide for Planners, Policymakers and Citizens* 220–21 (1998).

35. Gottlieb, *supra* note 32, at 19.

36. Judith Coburn, *Whose River Is It Anyway? More Concrete versus More Nature: The Battle over Flood Control on the Los Angeles River Is Really a Fight for Its Soul*, L.A. Times, Nov. 20, 1994, at 18–24, 48–54.

37. *See, e.g.*, Lewis MacAdams, *Restoring the Los Angeles River: A Forty-Year Art Project*, Whole Earth Review, Spring 1995, at 63.

38. *Id.*

39. *Id.*

40. *Id.*

41. *Id.*

42. Lewis MacAdams, *Sharing Memories of the L.A. River*, L.A. Times, Nov. 28, 1985, at 22.

43. Dick Roraback, *From Base Camp, the Final Assault Series: In Search of the L.A. River. Last in an Intermittent Series*, L.A. Times, at. *See also supra* notes 11, 17, 18, 20.

44. Gumprecht, *supra* note 26, at 127, 246. Soft bottom areas of the river are unpaved because the groundwater table was sufficiently high that pouring concrete on the bottom of these sections was not deemed viable by engineer managers.

45. *Id.*

46. *Id.*

47. Gottlieb, *supra* note 32, at 20.

48. *Id.*

49. Interview with Lewis MacAdams, founder, Friends of the L.A. River (Oct. 1998).

50. Gumprecht, *supra* note 26, at 273–74.

51. *Id.*

52. *Id.*

53. *Id.*

54. Gottlieb, *supra* note 32, at 20.

55. Roraback, *supra* note 11.

56. Orsi, *supra* note 6, at 148–52. *See also* Gumprecht, *supra* note 26, at 279.

57. Orsi, *supra* note 6, at 151.

58. *Id.*

59. *Id.*

60. *Id.*

61. *See* Robert Gottlieb, Mark Vallianatos, Regina Freer & Peter Dreier, *The Next Los Angeles: The Struggle for a Livable City* (2005).

62. Gumprecht, *supra* note 26, at 297–98.

63. *Id.* at 298–99.

64. Christopher Kroll, *Changing Views of the River*, California Coast and Ocean, Summer 1993, at 32.

65. Riley, *supra* note 34. FoLAR's notion of flood management, influenced in part by Riley, was distinguished from flood control and included plantings, spreading grounds (low-lying bankside lands specifically designed to accommodate and dissipate high-volume flows), and other strategies to slow the flow of the river.

66. Gumprecht, *supra* note 26, at 298.

67. *Id.*

68. *Id.*

69. *Id.* at 298.

70. Duke Helfand, *Controversial LA River Project OKd*, L.A. Times, Apr. 7, 1995, at 1.

71. Gumprecht, *supra* note 26, at 283.

72. *Id.*

73. Orsi, *supra* note 6, at 156.

74. Gumprecht, *supra* note 26 at 283; *see also* Kroll, *supra* note 63, at 26.

75. Lewis MacAdams, founder, Friends of the L.A. River, address at Occidental College Environment and Society Class (Oct. 28, 1998).

76. *Id.*

77. *Id.*

78. *Id.*

79. *Id.*

80. *Id.*

81. Occidental College, *Learning by Doing, available at* http://www.oxy.edu/ x676.xml (last visited Sept. 1, 2005).

82. Urban and Environmental Policy Institute, *available at* http://departments .oxy.edu/uepi/about/index.htm (last visited Sept. 1, 2005).

83. Gottlieb & Azuma, *supra* note 2, at 10–11, 6–9, 4–5, 24–25.

84. *Id.* at 12–14, 8–9, 25–26, 22.

85. Robert Gottlieb, Elizabeth Braker & Robin Craggs, *Expanding the Opportunities and Broadening the Constituency for Interdisciplinary Environmental Education*, Report to the Andrew W. Mellon Foundation, Apr. 15, 2003.

86. *Cornfield of Dreams: A Resource Guide of Facts, Issues and Principles, available at* http://www.sppsr.ucla.edu/dup/research/main.html, (last visited Sept. 1, 2005).

87. *Id.* at 19–20.

88. Robert Gottlieb, *Rediscovering the River*, Orion Afield, Spring 2002, at 32.

89. Lecture Series, Re-Envisioning the L.A. River (Oct. 1, 1999).

90. *Id.*

91. *Id.*

92. Gottlieb & Azuma, *supra* note 2, at 5.

93. Friends of the L.A. River, *The River through Downtown, available at* http://www.folar.org/about.html (last visited Jan. 19, 2005).

94. Personal communication from Chi Mui, L.A. Chinatown activist, to Robert Gottlieb (Dec. 11, 2001).

95. *Id.*

96. Paul Stanton Kibel, *Los Angeles' Cornfield: An Old Blueprint for New Greenspace*, 23 Stan. Envtl. L.J. 275, 308 (2004).

97. Gottlieb, *supra* note 88, at 31.

98. *Id.*

99. Lewis MacAdams had first learned about Majestic Realty's plans when he over-heard a conversation in City Council member Mike Hernandez's office to the effect that Majestic was moving quickly to obtain a Mitigated Negative Declaration to proceed with their warehouse plans. MacAdams immediately called environmental attorney Jan Chatten-Brown to delay the proceedings to give time for FoLar and other river advocates to mobilize around the issue. The discussion with Nichols and Marcus occurred soon after. Personal communication from Lewis MacAdams, founder of Friends of the L.A. River, to Robert Gottlieb (Dec. 11, 2001).

100. Meeting at Eagle Rock Restaurant (Oct. 1, 1999) (quote from memory of occasion, Robert Gottlieb).

101. Gottlieb, *supra* note 88, at 33.

102. Lewis MacAdams & Robert Gottlieb, *Changing a River's Course: A Green-belt versus Warehouses,* L.A. Times, Oct. 3, 1999, at M1, 6.

103. *Cornfield of Dreams, supra* note 86, at 114–1.

104. Gottlieb & Azuma, *supra* note 2, at 30.

105. Personal communication from Lewis MacAdams, founder, Friends of the L.A. River, to Robert Gottlieb (Sept. 14, 2000).

106. Kibel, *supra* note 96, at 325–30.

107. *Id.* at 330.

108. Gottlieb, *supra* note 88, at 33.

109. Orsi, *supra* note 6, at 156–5.

110. Gottlieb & Azuma, *supra* note 2, at 20.

111. *Uncommon Ground: Rethinking the Human Place in Nature* (William Cronon ed., 1996).

112. Gottlieb and Azuma, *supra* note 2, at 20.

113. D. J. Waldie, *Changing the River's Course in Pursuit of Public Spaces,* L.A. Times, Oct. 3, 1999, at 1.

114. Safe Neighborhood Parks, Clean Water, Clean Air, and Coastal Protection Bond Act (2000), *available at* http://primary2000.ss.ca.gov/returns/prop/00.htm (last visited Aug. 5, 2005); Water Security, Clean Drinking Water, Coastal and Beach Protection Act (2002), *available at* http://vote2002.ss.ca.gov/Returns/prop/00.htm (last visited Aug. 5, 2005).

115. Robert Gottlieb, *Expanding Environmental Horizons,* L.A. Times, Apr. 16, 2000, at 6.

116. Los Angeles Public Resources Code 32600-32602, *available at* http://leginfo.ca.gov/cgi-bin/displaycode?section=prc&group=32001-33000&file=32600-32602 (last visited Aug. 5, 2005).

117. Gottlieb & Azuma, *supra* note 2, at 15.

118. *Id.*

119. *Id.*

120. *Id.* at 13–15; *see also* Gottlieb et al., *supra* note 61, at 108.

121. Department of Landscape Architecture, Harvard University. Graduate School of Design, *L.A. River Studio Book* 21 (2002).

122. Gottlieb and Azuma, *supra* note 2, at 24.

123. *Id.*

124. *Id.*

125. *Id.*

126. Waldie, *supra* note 113.

127. *Id.*

128. D. J. Waldie, *As We Gather at the River,* L.A. Times, July 23, 2000, at 1.

129. *Id.*

130. D. J. Waldie, *Reclaiming a Lost River, Building a Community*, N.Y. Times, July 10, 2002, at A 21.

131. *Id.*

132. *Id.*

3

Bankside Washington, D.C.

Uwe Steven Brandes

Our Capital's Other River

For many decades, the Anacostia River—its shoreline, waterfront neighborhoods, and watershed—has been neglected by those responsible for its stewardship. The river's water is severely polluted, obsolete transportation infrastructure isolates neighborhoods and divides the District of Columbia (DC) into areas east and west of the river, public parks are underutilized and suffer from chronic disinvestment, and several communities along the river are among the poorest in the metropolitan Washington, D.C., region. With the river forming a boundary between race and class[1] and with over 70 percent of the river's lands in public ownership, the need to rethink the management of this urban river is clear. While the river can be understood only as a function of its watershed, the focus of this essay is on those lands within the District of Columbia, which form the last seven-mile stretch of river corridor before the confluence with the Potomac River.

Today, the effort to recapture the Anacostia follows in Washington's tradition of great public-works initiatives. The original plan for the city, now 200 years old, established the urban framework for a great national capital stretching between the Potomac and Anacostia Rivers.[2] One hundred years ago, the Senate Park Commission's McMillan Plan envisioned Washington's most memorable civic places along those rivers, including the National Mall and Rock Creek Park, but the McMillan Plan's vision of an ecological greensward along the Anacostia River was never realized.[3]

This essay explores the federal-local planning process known as the Anacostia Waterfront Initiative (AWI), which has produced a development vision for the Anacostia River and its neighborhoods that may prove as powerful and enduring as previous city-building endeavors that have shaped the nation's capital into what it is today.

The following guiding principles were established at the outset of the AWI in March 2000:[4]

· Create a lively urban waterfront for a world-class, international capital city;
· Produce a coordinated plan that can be implemented over time;
· Restore the Anacostia's water quality and enhance the river's natural beauty;

The Anacostia Watershed

Figure 3.1
Map of the Anacostia watershed, 2003. Reprinted with permission of the Anacostia Waterfront Corporation.

· Reconnect neighborhoods along the river and link their communities to the river;

· Link distinctive green parks, varied maritime activities, and unique public places into a continuous public realm;

· Embrace sustainable and low-impact development in waterfront neighborhoods;

· Stimulate economic development and job creation, ensuring that existing residents and low-income communities benefit and share in the redevelopment;

· Engage all segments of the community to foster river and watershed stewardship;

· Address issues and concerns raised by the community; and

· Promote excellence in architectural and landscape design in all aspects of the endeavor.

Before exploring the development of these goals, it is useful to place the river in its broader context.

The River Corridor

Environment and Geography

The Anacostia River forms a tributary to the Potomac River that drains 176 square miles of land in Maryland (83 percent) and the District of Columbia (17 percent).[5] It flows for seven miles through Washington on the eastern side of the city.[6]

The river's watershed is the most densely populated subwatershed in the Chesapeake Bay and has been identified as one of the bay's three primary toxic hotspots.[7] The river's water quality has been described as one of the most endangered in the nation.[8] Primary sources of contamination are (1) *legacy toxins* concentrated in the silt at the bottom of the river, (2) *nonpoint-source contaminants* born in urban stormwater runoff throughout the watershed, and (3) *discharges* of sanitary sewage and of combined stormwater and sanitary sewage that overflow into the river during an average of over seventy-five "events" per year.[9]

Within DC, the shoreline is overwhelmingly owned by the federal government.[10] Major facilities include the National Arboretum, the National Park Service's Anacostia Park, the Washington Navy Yard, and the United States Army's Ft. McNair.[11] The District of Columbia leases or has operational control over several federal parcels, including RFK

Stadium, District of Columbia General Hospital, the District of Columbia Jail, the Main Sewage Pump Station, and all of the streets and bridges that form the city's transportation system.[12] The District of Columbia also owns several sites, including the southwest waterfront. Two former electricity power plants along the river are owned by the Potomac Electric Power Company.[13] In total, over 90 percent of the river's shoreline is in public ownership.[14]

History and Evolution

The river—initially the commercial lifeline of Washington and the upstream port of Bladensburg, Maryland—already had been severely compromised by erosion and siltation by the time of the Civil War.[15] During the nineteenth century, weapons manufacturing and ship-building activities at the Navy Yard provided enough jobs to encourage the first residential community on the east side of the river, originally named Uniontown and today referred to as Historic Anacostia.[16]

When the United States Army Corps of Engineers began implementing the vision of the McMillan Plan in the 1910s and 1920s, hundreds of acres of tidal estuary were filled, and the river's configuration was re-engineered. But the proposed damming of the river at the propsed new river crossing at Massachusetts Avenue SE proved infeasible and was never implemented.[17] The highway building era of the 1950s took advantage of the extensive areas of reclaimed wetlands to construct new regional infrastructure.[18] The newly created lands along the river were eventually transferred to the Department of Interior with the designation of park use, but with the land criss-crossed by regional infrastructure, the great park-building effort envisioned by the McMillan Plan never came to be.[19]

By the mid-twentieth century, the neighborhoods along the river became one of the primary targets of Washington's urban-renewal actions, in which existing residences and businesses were designated slums under federal law, razed on a block-by-block basis, and replaced with new public or publicly assisted housing projects.[20] The southwest quadrant of the city—Washington's longtime commercial harbor district—became one of the nation's largest urban-renewal projects. In fact, the redevelopment of this neighborhood was the subject of the 1954 landmark United States Supreme Court decision in *Berman v. Parker*, which upheld the municipal powers of eminent domain for purposes of urban redevelopment.[21] Many residents, generally poor and African American, were relocated into neighborhoods farther east, resulting in a concentration of public housing along the river and a legacy of social and neighborhood disruption that lives on to today.[22]

Demographics and Economy

Today, residential neighborhoods abut the federal lands along the river, although almost all of them lack any or easy access to the river.[23] Historic neighborhoods include Capitol Hill, Fairlawn, and Historic Anacostia.[24] Several neighborhoods along the river were developed or redeveloped during the urban-renewal era between 1950 and 1970. These include the Southwest Waterfront, the Near Southeast, River Terrace, Mayfair Mansions, and Carver Langston.[25] The character of several of these neighborhoods is defined by large concentrations of public housing constructed in a low-rise barracks style. While it is hard to imagine today, the construction of postwar housing in the District of Columbia often occurred on farmland that was still in agricultural production into the 1950s, only a few miles from the Capitol building.[26] Settlements on the east side of the river were referred to in planning and urban-renewal documents as "rural blight."[27]

Neighborhoods along the river are home to some of the poorest residents of the city and the region, with the average per capita income averaging less than half that of the region and with concentrations of poverty in select neighborhoods approaching one in four households.[28] Two of the city's eight wards are located east of the river, with demographics of race approaching 95 percent African American.[29]

While the real estate market has been on a steep upswing since the late 1990s, large-scale development prior to 2000 was limited largely to the downtown.[30] Metropolitan Washington saw significant suburban growth in the 1980s and 1990s, but it is only since the late 1990s that significant new residential development has been undertaken within DC.[31] With the Washington metropolitan region now considered one of the strongest real estate markets in the country[32] and with building heights within the District of Columbia regulated by an act of Congress,[33] the city's downtown core must grow to the east toward several large, underutilized tracks of land along the river.[34]

The Recent Legal and Administrative Context

Several legal actions have defined the recent history of the river. Citizen and nonprofit organizations have dramatically influenced several large public-works projects, including the proposed construction of an amusement park[35] and the planned extension of a freeway across the river.[36] Using the Clean Water Act, several nonprofit organizations have pursued litigation regarding the combined sewage overflows.[37] The District of Columbia's Water and Sewer Authority has recently formulated a strategy to bring the city's sewer infrastructure into compliance with Federal Environmental Protection Agency (EPA) standards at an estimated cost of over $1.3 billion over the next twenty years.[38]

The local political context, as reflected in these actions and largely defined in the 1990s by the takeover of DC finances by the congressionally legislated Control Board, became the backdrop for the 1998 election of Anthony A. Williams as mayor of the District of Columbia.[39] Building on his personal interest in ecology and rivers and his political commitment to social justice in neighborhoods throughout the city, Williams has raised the challenges associated with river to the highest level of his attention.[40]

In 2000, Williams successfully forged a partnership between the city government and the federal agencies that owned land along the river.[41] Conceived as the Anacostia Waterfront Initiative, the partnership was memorialized in a memorandum of understanding (MOU) that was signed by the mayor and over a dozen federal agencies in March 2000 at the Navy Yard.[42] The Anacostia Waterfront Initiative joined the District of Columbia and federal agencies in a participatory planning process to form a common policy and development vision for the river and its public lands.[43] This process, unprecedented for the Anacostia River and unprecedented in the history of urban planning in the District of Columbia, was described by Williams as one of the most important partnerships ever created between DC and the federal government.[44]

Planning Process

In addition to establishing guiding principles, the memorandum of understanding contains a number of innovative provisions that made the AWI an unprecedented planning process in the history of Washington.[45]

First, by identifying the District of Columbia's Office of Planning as the lead agency in the process, the city was put in a leadership role to coordinate the vision for the river's waterfront lands, including the federal lands. Second, it established a joint steering committee comprised of Office of Planning, the National Park Service, and the General Services Administration to oversee the progress of the planning. Third, it established a mandate to engage the citizens of the District of Columbia in the planning process.[46]

In consultation with City Council members, the Office of Planning established a 150-person Citizens Steering Committee that included opinion leaders representing individual neighborhoods, environmental advocacy groups, and the business, city-planning, and architectural-design community.[47] This committee was formed to build support for the planning process as well as to create a forum to discuss the major public-policy issues related to the river.[48] Concurrent to the quarterly meetings of the steering committee, the Office of Planning sponsored over thirty community workshops and focus-group sessions in six neighborhood target areas.[49] By the end of 2004, over 5,000 individuals attended these neighborhood workshops or attended the well-publicized progress presentations held at the National Building Museum or at the Arena Stage theater located at the Southwest waterfront.[50]

AWI Framework Plan

Growing out of the dialogue fostered between citizens and the federal agencies, the Office of Planning produced the Waterfront Framework Plan to guide the river's redevelopment over the course of the next generation.[51] To achieve the goal of a great waterfront along the Anacostia River, the Framework Plan identifies five planning themes, which form the basis for the five chapters of the plan.[52] Each of these themes responds to citizen concerns or public policy debates focused on the river corridor.[53]

A Clean and Active River (Environment)   Community and environmental advocates maintained that the river be restored to "fishable and swimmable" levels of water quality.[54] This was and continues to be the most discussed recommendation of the Framework Plan, given that the amount of public funding necessary to implement the Sewer Long-Term Control

Plan as well as watershed restoration totals nearly $2 billion.[55] Furthermore, given the significance of ongoing nonpoint-source contaminant loading and the location of the majority of the river's watershed in Maryland, the Framework Plan highlighted DC's political predicament—it is down stream—and has no way on its own to force the State of Maryland to prioritize this watershed-restoration effort.[56]

Eliminating Barriers and Gaining Access (Transportation)   While neighborhood groups had recently succeeded in their efforts to halt the expansion of the city's freeway network on the Anacostia waterfront, few stakeholders offered a positive vision for the future of traffic around the river.[57] One issue that the planning process helped articulate was that the primary transportation barrier was not the river itself but rather was the poorly designed freeways that were constructed parallel to it to carry workers from the downtown to the newly emerging suburbs of metropolitan Washington.[58]

Figure 3.2
Sketch of a new Anacostia riverfront from the Anacostia Waterfront Initiative's *Framework Plan*, 2003. Reprinted with permission of the District of Columbia Office of Planning.

A Great Riverfront Park System (Public Realm)   The steering committee played an important role in elevating the discourse on parks and advocated for design and environmental excellence to match the standards of other parks in the capital and in other cities.[59] From the outset of the planning process, the mayor championed the idea of a continuous Riverwalk on both sides of the river.[60] The Riverwalk captured the public and media's imagination and helped convince the District government to fund several demonstration segments of the Riverwalk, which made the notion of continuous public access to the river a concrete and widely accepted goal.[61]

Cultural Destinations of Distinct Character (Culture and Institutions)
The waterfront initiative was preceded by a citywide Museum and Memorials Plan completed by the National Capital Planning Commission (NCPC).[62] Given that potential sites for additional memorials on the National Mall are scarce, NCPC completed this plan to highlight opportunities to locate new monuments off the Mall.[63] Waterfront sites represented many of the most promising locations.[64] The Framework Plan sought strategies whereby new memorials would reinforce existing river attractions, as well as existing, underappreciated historic resources.[65] Recently, the District's efforts to construct a new ballpark for the Washington Nationals baseball team is part of the strategy to transform a segment of the river into a citywide and regional destination that elevates the river's civic importance.[66]

Building Strong Waterfront Neighborhoods (Economic Development)
As the planning process proceeded, the issue of residential gentrification and potential resident displacement was debated even more passionately than the need to restore the river's environmental quality. The gentrification debate was made more complex by a series of broadly discussed papers (written by Alice Rivlin) that argued that the fiscal health of the District of Columbia was dependent on an economic development strategy that increased the city's population by at least 100,000 persons.[67] Ultimately, the Framework Plan recommended adding 15,000 new units of housing along the river, justified by the opportunity to grow mixed-income neighborhoods without displacing existing residents.[68]

The Center of Twenty-First-Century Growth in Washington[69]

With Washington's downtown nearly built out, the city's pattern of growth is moving steadily eastward toward and across the Anacostia River.[70] The capacity of Washington to grow is now inextricably linked to recentering its growth in the coming decades around the Anacostia River.[71] The Anacostia's neglected parks and natural environment are being reexamined as the locus of a new civic space in Washington, which citizens from all over the city might find attractive.[72]

The recovery of the Anacostia Waterfront has the potential to reunite the capital city economically, physically, and socially.[73] It could reinvigorate the river with new resident stewards, reclaim the waterfront's parklands for community use, reconnect neighborhoods with new bridges and roads, create new museums and monuments, and expand opportunities to live, work, play, and learn in an urban setting.[74] The vision for the Anacostia is one of vibrant and diverse settings for people to meet, relax, encounter nature, and experience the heritage of Washington.[75] The Anacostia Waterfront Initiative seeks to ensure that the social and economic benefits derived from a revitalized waterfront are shared in an equitable fashion by those neighborhoods and people for whom the river has been distant, out-of-reach, or unusable.[76]

Planning at the Neighborhood Scale

While the Framework Plan explores riverwide issues, target-area plans were prepared to chart redevelopment strategies on a neighborhood scale.[77] Six target-area plans apply the five waterfront planning themes to a site-specific context.[78] Each was completed with direct involvement of community stakeholders and then brought to the City Council for approval as a supplement to the city's comprehensive plan.[79]

Each target-area plan opened planning issues specific to its neighborhood that were resolved in the context of riverwide goals outlined in the Framework.[80] Conflicts and tradeoffs between riverwide goals and local plans required balance, with each neighborhood expressing its own set of challenges. Height and density impacts of proposed high-density development were most pronounced at the Southwest Waterfront, where existing residents were likely to have river views marred by new buildings.[81] Concerns over housing affordability and the management of public-housing assets were most pronounced in the Near Southeast, where the planning process included a public-private development proposal to redevelop the Capper Carrolsburg housing project. The issues of proposed land uses were most pronounced at Hilleast, where the District had recently closed the public hospital and where the need to accommodate municipal services, such as healthcare clinics and correctional uses, was balanced with the expansion of the residential uses to connect Capitol Hill with the river. The competing open-space objectives of recreation versus ecological restoration were part of the debate over the Anacostia Park.[82]

In summary, issues of environmental restoration and gentrification were discussed on a citywide scale, while neighborhood quality-of-life issues like parks, traffic, and retail development were advocated for on a neighborhood-by-neighborhood basis.

Waterfront Investment during the Planning Process

The planning endeavor became more dynamic with several real-time public investments.[83] While these projects tended to politicize individual advocacy groups, they also made plausible the notion of a reenergized and transformed river corridor and kept the Anacostia Waterfront Initiative in the pages of the *Washington Post*.[84] The U.S. Navy, which was responsible for the oldest continuously operating Navy Yard in the country, played a lead role in this reinvestment through its efforts at its waterfront facilities by consolidating regional employment at the Yard through the Base Realignment and Closure Act.[85] Over $400 million was invested in rehabilitating industrial buildings listed on the National Register of Historic Places into high-tech navy office space, and employment nearly tripled to almost 11,000 enlisted and civilian employees.[86]

Several city agencies mobilized to demonstrate "immediate impact." The city's Watershed Protection Division, working in a joint venture with the United States Army Corps of Engineers, reconstructed over forty acres of wetlands along the river.[87] The newly formed Water and Sewer Authority made interim investments in inflatable dams within the sewer system to curb combined sewer overflow discharge into the river by 23 percent.[88]

Mayor Williams initiated a series of high-profile public-private partnerships utilizing newly legislated tools, such as tax increment financing

(TIF) and payment in lieu of taxes (PILOT).[89] The redevelopment of the Capper Carrollsburg public-housing complex was perhaps the most innovative and highly leveraged housing project to be completed under HUD's HOPE VI program. With the mayor guaranteeing a one to one replacement of all public housing, the project increased land densities to double the amount of housing units by supplementing 700 units of public housing with 400 units of subsidized housing and 400 units of market-rate housing.[90] Federal agencies responded as well, with the General Services Agency playing a key role by selecting a river site for the new headquarters for the United States Department of Transportation and by disposing excess land to the private sector under a special act of Congress.[91]

Undoubtedly the most high-profile public commitment the mayor made was to aggressively pursue the construction of a new baseball ballpark along the river. With public finance legislation that made national headlines in December 2004, the Washington Nationals began to call the Anacostia their home as of April 2005 as they played in RFK Stadium and will call it their permanent home when the new ballpark is delivered directly south of the Capitol only one block from the river. This $611,000,000 public investment has put the Anacostia at the center of regional and national attention.

The private sector responded—and is responding—to these public investments with an initial wave of construction that included several new commercial office buildings and hundreds of new units of housing with thousands more in the predevelopment stage.[92] Waterfront planning events were well attended by members of the real estate development community, and the perception of the Anacostia River changed dramatically in the press and among several local professional associations that ultimately championed the AWI.[93]

The Anacostia Waterfront Corporation

The Framework Plan proposed the creation of a new dedicated entity to oversee the transformation of the river corridor.[94] Currently, lands along the river fall under the jurisdiction of multiple federal and local authorities and agencies, not one of which has a clear mandate for revitalizing the waterfront.[95] It was recognized that a new institution would help ensure that the resources necessary to implement the plan are advocated for and wisely and equitably invested for the river as a whole.[96] The role of this new entity—which became known as the Anacostia Waterfront Corporation—would be to oversee implementation of the plan, ensure sustained public participation by acting as a design "clearing house," and be responsible for promoting waterfront activities and in some cases managing public spaces.[97]

In considering how the Anacostia Waterfront Corporation would be organized, several models were explored based on federal-local actions in other cities. Among those evaluated were the Pennsylvania Avenue Development Corporation[98] in Washington, D.C., Presidio Trust in San Francisco,[99] and the Southern Nevada Land Act.[100] Each of these redevelopment projects had been initiated with federal legislation, and each had significant localized outcomes as its purpose. In the case of the Southern Nevada Land Act, the proceeds from federal land disposition are reinvested into federal lands but in a partnership arrangement with local jurisdictions.[101]

Ultimately, the existing federal legislation models were dismissed for three primary reasons. First, the tools to creatively finance public-private partnerships resided with the District of Columbia. Second, given that the District continues to be under the oversight of the Congress, the opportunity for a unique federal partnership was de facto in place. Third, the initiative itself had always been focused on reconnecting the citizens of the District to their river. A locally chartered organization (government-sponsored enterprise) appeared to be the most appropriate and promising vehicle to transform the river into a public asset and amenity. In summary, the structure that emerged took advantage of the city's own powers of creative financing but formed a semiautonomous municipal entity with which land-owning federal agencies and the Congress could easily partner.

Anacostia Waterfront Corporation Act

The District of Columbia Anacostia Waterfront Corporation Act was passed by the City Council in 2004 and creates a District government-chartered corporation charged with the development, promotion, and revitalization of the Anacostia River waterfront.[102] With a board that includes both mayoral appointees as well as ex-officio members representing key District and federal agencies,[103] the Corporation is a city-sponsored entity poised to become a development partner for both municipal and federal agencies as well as the private sector. Other cities—such as London, San Francisco, Barcelona, and Pittsburg—have demonstrated that successful waterfront development often hinges on the role of a single-purpose, dedicated public entity and on strategic coordination between many government agencies (often involving state, municipal, and federal jurisdictions over long periods of time) to complete projects that have physical challenges unique to waterfronts.[104] Asking an existing government agency to "do it all" runs counter to almost every other city in the nation that has decided to implement an aggressive waterfront-development program.

A single development corporation appeared as the most viable structure to make sure that all the various components of the Anacostia Waterfront Initiative—residential development, maritime uses, recreational uses, transportation infrastructure, commercial and retail development, cultural uses, and environmental restoration—were coordinated in a way that maximizes the benefit of the river as a natural asset to the District of Columbia.

Conclusion

The Anacostia Waterfront Initiative represents one of the most important partnerships between local and federal agencies in the District of Columbia. It is unprecedented in the history of urban planning in Washington due to its inclusion of neighborhoods on both sides of the river and its multidisciplinary approach to environmental restoration. It is the first participatory planning process conducted in the District of Columbia that was explicitly conceived of as a local-federal partnership to plan for local and federal lands simultaneously.

Recent public actions seek to institutionalize the spirit of the planning partnership by forming a dedicated entity with a single purpose of realizing the AWI Framework Plan and with a governance structure that includes both local and federal representation. The goal of the AWI is nothing short of the rebirth—ecological, economic, social, and cultural—of the Anacostia River in a manner that responds to contemporary urban and environmental dilemmas, while following in the footsteps of the great city-building traditions of the nation's capital city.

Notes

1. District of Columbia Office of Planning, *2000 Population by Single Race and Hispanic Origin by Ward, available at* http://www.planning.dc.gov/planning/cwp/view,a,1282,q,569460.asp (last visited Aug. 5, 2005) (providing data that show that demographics of the two city wards east of the river are 96.8 percent and 92.4 percent African American).

2. Ruth W. Spiegel, *Worthy of the Nation: The History of Planning for the National Capital,* National Capital Planning Commission Historical Studies 19 (1977), *citing* the *L'Enfant Plan for Washington* (1791). The L'Enfant plan staked out key public tracts of land along the river for diverse uses such as markets, hospitals, and military installations. It is not widely known that the plan and the rights-of-way established by the plan are listed on the National Register of Historic Places. Because of this listing, virtually all improvements and alterations to the street and block pattern along the river corridor are subject to historic-preservation review.

3. *Id.* at 118–36. *The Improvement of the Park System of the District of Columbia,* prepared by the U.S. Senate, Committee on the District of Columbia, engaged leading design practitioners of the day, including Daniel Burnham, Charles McKim, Frederic Law Olmstead Jr., and Charles Moore. The Anacostia was envisioned as a vast water park in its northern reaches and an urbanized quay along its southern reaches near the Navy Yard.

4. District of Columbia Office of Planning, *Anacostia Waterfront Initiative Framework Plan* 11 (Nov. 2003).

5. District of Columbia Water and Sewer Authority, *WASA's Recommended Combined Sewer System Long-Term Control Plan* 2-2 (2002).

6. Author's geographic research.

7. Chesapeake Bay Program, *Targeting Toxics: A Characterization Report* (June 1999).

8. American Rivers, *America's Most Endangered Rivers List of 1993* (1993), *available at* http://www.americanrivers.org/site/pagerserver?pagenname=AMR_content_97b0 (last visited in 1993).

9. District of Columbia, *WASA, supra* note 5, at 3–4.

10. District of Columbia, *Anacostia Waterfront, supra* note 4, at 16.

11. *Id.* at 10.

12. *Id.* at 10.

13. *Id.* at 17.

14. *Id.* at 16.

15. *Id.* at 14.

16. Spiegel, *supra* note 2, at 58.

17. *Id.* at 142–43.

18. *Id.* at 281.

19. District of Columbia, *Anacostia Waterfront, supra* note 4, at 14.

20. Spiegel, *supra* note 2, at 318.

21. *Berman v. Parker*, 348 U.S. 26, 33 (1954).

22. Personal interviews by the author with residents of Arthur Capper Carrolsburg Dwellings during Hope VI planning workshops in Washington, D.C. (July–Oct. 2002). Several elderly citizen stakeholders in the Anacostia Watershed Initiative planning process who are current residents of public housing traced their personal and family history to the southwest waterfront neighborhood from which they were relocated by the Redevelopment Land Agency. Throughout AWI planning charrettes and workshops, the urban-renewal era of city planning in Washington was colloquially referred to by many citizen stakeholders as "negro removal."

23. District of Columbia, *Anacostia Waterfront, supra* note 4, at 96.

24. *Id.* at 97.

25. *Id.* at 97.

26. Spiegel, *supra* note 2, at 237.

27. *Id.* at 237.

28. District of Columbia, *2000 Population, supra*, note 1.

29. *Id.*

30. *Id.*

31. *Id.*

32. Urban Land Institute and Price Waterhouse Coopers, *Emerging Trends in Real Estate 2004* 31–32 (2004).

33. The Building Heights Act of 1910, 36 Stat. 452, 455 (1910).

34. District of Columbia, *Anacostia Waterfront, supra* note 4, at 9.

35. *Anacostia Watershed Society v. Babbitt*, 871 F. Supp. 475 (D.D.C. 1994).

36. *District of Columbia Federation of Civic Associations v. Airis*, 391 F.2d 478 (D.C. Cir. 1968).

37. *Kingman Park Civic Association v. U.S. Environmental Protection Agency*, 84 F. Supp. 2d 1 (D.D.C. 1999).

38. District of Columbia, *WASA, supra* note 5, at 9.

39. District of Columbia, *Anacostia Waterfront, supra* note 4, at 3.

40. *Id.* at 3. An avid canoeist and amateur ornithologist, Williams kicked off his campaign on the Anacostia's Kingman Island, symbolic through its location in the middle of the river and a location that has been off limits to public access for decades.

41. *Id.* at 4.

42. *Id.* at 8.

43. *Id.* at 4.

44. *Id.* at 3.

45. *Id.* at 4.

46. *Id.* at 4.

47. *Id.* at 131.

48. *Id.* at 11.

49. *Id.* at 130.

50. District of Columbia, Office of Planning, Stakeholder attendance records.

51. District of Columbia, *Anacostia Waterfront, supra* note 4, at 8.

52. *Id.* at 21.

53. *Id.* at 21.

54. *Id.* at 23.

55. *Id.* at 21.

56. *Id.* at 26.

57. *Id.* at 37.

58. *Id.* at 37.

59. *Id.* at 59.

60. *Id.* at 60.

61. *A River on the Rise,* Washington Post, Apr. 3, 2003.

62. National Capital Planning Commission, *Memorials and Museums Master Plan* (2001).

63. *Id.*

64. *Id.*

65. District of Columbia, *Anacostia Waterfront, supra* note 4, at 80.

66. For a discussion of the site-analysis guidelines that were used to complete the site selection of the ballpark, *see* District of Columbia Sports and Entertainment Commission, Deputy Mayor for Planning and Economic Development, and Washington Baseball Club, *Washington, D.C., Major League Baseball Park Site Evaluation Project* (Nov. 6, 2002).

67. Alice M. Rivlin, *Revitalizing Washington's Neighborhoods: A Vision Takes Shape* (2003).

68. District of Columbia, *Anacostia Waterfront, supra* note 4, at 17.

69. These two paragraphs represent a synopsis of a general public-information overview published by the District of Columbia Office of Planning intended to communicate the significance of the *Anacostia Waterfront Framework Plan* to a broad, general audience. The brochure is entitled *The Anacostia Waterfront: Imagine, Act Transform* and was also accompanied by a DVD format animation.

70. District of Columbia, *Anacostia Waterfront, supra* note 4, at 17.

71. *Id.*

72. *Id.* at 11.

73. *Id.* at 8–9.

74. *Id.* at 10.

75. *Id.* at 10–11.

76. *Id.* at 10–11.

77. *Id.* at 107.

78. *Id.* at 109.

79. Each target-area plan may be found at www.anacostiawaterfront.net.

80. District of Columbia, *Anacostia Waterfront, supra* note 4, at 109.

81. District of Columbia, Office of Planning, *Southwest Waterfront Development Plan* 122 (2002).

82. District of Columbia, *Anacostia Waterfront, supra* note 4, at 109.

83. It is important to note that the Anacostia Waterfront Initiative memorandum of understanding did not conceive of the Initiative as only a planning effort. Significant advocacy efforts, the revision of key zoning regulations, positive press, and positive District-federal interagency coordination resulted in over $125 million of public appropriations and over $1.5 billion of private investment during the course of the planning process itself.

84. The press coverage of the Anacostia Waterfront Initiative in the *Washington Post* was extensive. Over a three-year period, dozens of articles appeared, many of them features on the cover of the *Post*'s Metro section, ensuring that the project became understood as a citywide endeavor. Select *Washington Post* articles include *On the Waterfront,* Nov. 25, 2000, at G01; *Shaping the City*, Feb. 10, 2001, at G03; *Hope on the Waterfront,* Apr. 20, 2001, at A4; *Making a Case for Capital's Other River*, May 17, 2001, at DZ10; *Want to Save the Anacostia?*, June 21, 2001, at D01; *Lively—Costly—Area Envisioned along the Anacostia,* Nov. 9, 2001, at B01; *D.C. Backs Concept for Southwest Waterfront,* Oct. 8, 2003, at B01; *Anacostia Plan Wins Backing,* Jan. 16, 2004, at D01; *River of Dreams,* Jan. 17, 2004, at C01; *Neighborhoods Have a Big Role in Anacostia Waterfront Plan,* Jan. 19, 2004, at D01; *A Building Plan Runs through It,* Jan. 23, 2004; *Anacostia River's Dirty Little Secret,* Jan. 29, 2004, at B01.

85. Interview with Admiral Jan Gaudio, Naval District Washington (Mar. 30, 2005).

86. *Id.*

87. District of Columbia, *Anacostia Waterfront, supra* note 4, at 111.

88. CSO Update, District of Columbia, *Water and Sewer Authority* (2004).

89. *See* Deputy Mayor for Planning and Economic Development, *available at* http://www.dcbiz.dc.gov/dmped/cwp/view,a,1365,q,569383,dmpedNav, |33026||33028|.asp (last visited Aug. 5, 2005).

90. District of Columbia Housing Authority, *HUD Application for Federal Assistance, Summary Letter* (June 22, 2001). The Capper Carrolsburg project introduces market-rate units and thereby creates the economics that allow all public units to be replaced in kind with no net loss of public-housing units. This strategy was a direct response to public-housing resident concerns voiced during

a waterfront planning workshop in May 2001, as documented by the Office of Planning summary brochure issued in the summer of 2001.

91. The Southeast Federal Center Public-Private Development Act, Pub. L. No. 106–407 (2000).

92. Interviews with developers confirmed that all tenants were defense contractors doing business with the U.S. Navy. Interview with Paul Robertson, Spaulding & Slye (Nov. 1, 2004).

93. The District of Columbia Building Industry Association, the Greater Washington Board of Trade, the District of Columbia Chamber of Commerce, and the Federal City Council all became important advocates for the plan and testified in support of its creation at the District of Columbia Council public hearing on February 11, 2004, (author's record). At the time the Framework Plan was adopted by the District of Columbia Council, the *Washington Post* ran a week-long series of front-page articles written by architecture critic Benjamin Forgey: *The Ripple Effect*, July 12, 2004, at A1; *Coming Clean about the Future*, July 13, 2004, at A1; *A Vision for the Southwest*, July 14, 2004, at A1; *Betting Big on Near Southeast*, July 15, 2004, at A1; and *Popularizing Poplar Point*, July 16, 2004, at D1.

94. District of Columbia, *Anacostia Waterfront, supra* note 4, at 124–25.

95. The National Capital Revitalization Corporation (www.ncrcdc.com), the Water and Sewer Authority (www.dcwasa.com), the District of Columbia Housing Authority (www.dchousing.org), and the District of Columbia Sports and Entertainment Commission (www.dcsec.com) are all purpose-created instruments of the District of Columbia, which have a significant stake in the Anacostia River.

96. District of Columbia, *Anacostia Waterfront, supra* note 4, at 124–25.

97. *Id.*

98. The Pennsylvania Avenue Development Corporation Act of 1972, Pub. L. No. 92–578 (1972).

99. The Presidio Trust Act of 1996, Pub. L. No. 104–333 (1996).

100. The Southern Nevada Public Land Management Act of 1998, Pub. L. No. 105–263 (1998).

101. *Id.*

102. Council of the District of Columbia, The Anacostia Waterfront Corporation Act of 2004, 15–616 (2004).

103. *Id.*

104. Urban Land Institute, *Advisory Services Report Anacostia Waterfront* 11 (2004).

4

Bankside Chicago

Christopher Theriot and Kelly Tzoumis

The Chicago River first attracted the Miami Indians, a branch of the Illiniwek, who set up camp near the mouth of a small river that flowed into Lake Michigan. They called their village Che-cau-gou after the scent of the wild onions that grew so prevalently in the area.[1] The exponential growth of the City of Chicago, from a small trading outpost to the second largest city in the United States, has transformed the original river ecosystem. Over the past two centuries, Chicago has employed a series of engineered technologies in its ongoing struggle to manage wastewater and provide clean drinking water. These engineered technologies have, in turn, created a new set of complex environmental and ecological responses—responses that have led to yet new proposals for technological fixes.

In this chapter, we begin with a brief overview of early engineered interventions in the Chicago River watershed, including the decision to reverse the river so that it flowed toward the Mississippi River basin rather than Lake Michigan. We then consider the origins and effects of two new far-reaching initiatives: the Tunnel and Reservoir Project and the electric fish barrier.

These initiatives are considered in the larger context of recent efforts to reenvision the Chicago River as an environmental amenity rather than a depository for contaminated urban runoff. In this reenvisioning process, the role of technology is a subject that merits closer attention. In terms of urban river restoration, technological responses can (depending on one's perspective) provide a means either to address or to avoid the root causes of ecological degradation.

Early Engineered Changes

The city of Chicago rose out of a flat plain on the banks of Lake Michigan where the soils were saturated and the water table was high. The level geography meant that the early sewage drainage ditches did not readily flow into the Chicago River and often sat stagnant. To remedy this situation, the Chicago City Council passed a resolution in 1885 that required the city streets and buildings to be raised between four to seven feet.[2] Beneath the new foundations of the elevated structures, the city's sewage ditches flowed more freely into the river and then the lake.[3]

However, this early engineered intervention failed to provide a permanent solution to the ever-increasing amount of human and animal wastes or to ensure an adequate supply of clean drinking water. One of the main sources of wastes was the Union Stock Yards, located along the South Fork of the Chicago River and made infamous in Upton Sinclair's 1905 novel *The Jungle*. As explained in Libby Hill's book *The Chicago River: A Natural and Unnatural History*:

The river was the reason for locating the [stock] yards there. Blood, guts, and other waste emptied into the lower portion of the South Fork. From there, the offal oozed into the South Branch, but it could not take enough of the odor away from the main part of the city. . . . Nothing as yet devised by the city sewage control could help the South Fork, as most of its flow was not water. As the offal settled to the bottom it began to rot. Grease separated and rose to the surface. Bubbles of methane formed on the bed of the river and rose to the surface, which was coated with grease. Some of these bubbles were quite large and when they burst a stink arose. There were many names for this part of the river, most unprintable. The one that stuck was bubbly creek.[4]

As the city grew, the population regularly suffered from reoccurring outbreaks of typhoid, dysentery, and cholera. Therefore, city planners decided to pump drinking water from further out in Lake Michigan in areas more removed from the contaminated river water that flowed eastward into the lake. In 1867, Irish immigrants completed a long tunnel sixty feet under the lake.[5] This tunnel was five feet in diameter, and its walls were lined with brick.[6]

In August 1885, a heavy rainstorm pushed contaminated wastewater from the Chicago River farther out into Lake Michigan, so that it reached the point of intake for the city's water supply. It became evident that the tunnel to deeper water would not provide a long-term solution for Chicago's freshwater needs. These shortcomings prompted the Illinois state legislature to create the Metropolitan Sanitary District of Greater Chicago (Chicago Sanitary District) in 1889.[7] The state granted the Chicago Sanitary District's elected commissioners broad enforcement, legislative, and taxing powers and charged the Chicago Sanitary District to both manage the city's wastewater and ensure a clean, safe supply of drinking water. Within several years, Chicago Sanitary District engineers developed plans for the most far-reaching engineered intervention to date—to redirect the Chicago River away from Lake Michigan and to send the city's wastewater downstream to St. Louis via a new canal. This proposed reversal of the river's flow was deemed a more economically and technically feasible solution than reducing the amount of human, animal, and industrial waste making its way into the Chicago River. This canal would connect the Chicago River to Mississippi River through the Des Plaines and Illinois rivers. For this massive public-works venture, workers dug a twenty-eight-mile canal starting at the southern branch of the Chicago River all the way to Lockport. At a depth of twenty-five feet, with a width exceeding 300 feet in some sections, the Ship Canal was one of the country's most ambitious engineered feats of its day.

Not surprisingly, the city of St. Louis, the State of Missouri, and communities upriver (soon to be downriver) of Chicago bitterly opposed becoming the recipient of Chicago's wastewater. The State of Missouri's legal challenge to the Ship Canal eventually landed in the United States Supreme Court in the case of *Missouri v. Illinois*. Fearful of an adverse ruling and the possible issuance of an injunction just as the Ship Canal was nearing completion, Chicago officials took matters into their own hands. At dawn on January 2, 1900, workers blew up the temporary dam separating the Chicago River from the Ship Canal as a group of Chicago Sanitary District commissioners looked on.

In the years following the opening of the Ship Canal, there were predictably adverse impacts on the bankside communities west of Chicago that were now the recipients of the city's untreated waste. In 1911, two biologists from the Illinois Natural History Survey studied the stretch just west of Chicago River and Illinois River confluence and reported "septic conditions" in which the water was "grayish and sloppy, with foul privy odors distinguishable in hot weather.... Putrescent masses of soft,

Chicago Waterway System

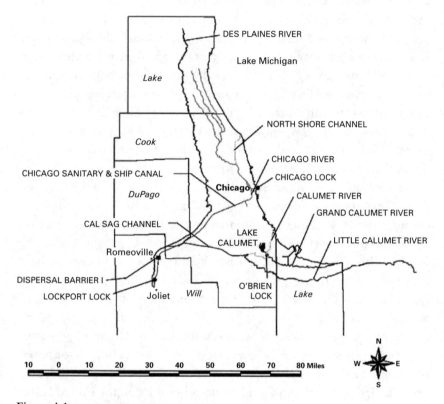

Figure 4.1
Map of the Chicago waterway system, March 2004. Reprinted with permission of the United States Army Corps of Engineers.

grayish, or blackish slimy matter, loosely held together by threads of fungi . . . were floating down the stream."[8]

Moreover, even the Ship Canal failed to provide a lasting solution for Chicago's wastewater problem. Frequent storm events forced the city to open the locks that separated the Chicago River from Lake Michigan to prevent flooding. The combined sewage system (sewage and stormwater) could handle 2 billion gallons per day, but a single rainstorm could overload the system with up to 5 billion gallons of runoff.[9] The area impacted by this combined sewer area covered approximately 375 square miles

with over 400 locations where overflows occurred in the city and surrounding suburbs.[10] By the early 1970s, sewage overflows into Lake Michigan occurred as often as once a week.

When it became clear that Chicago's existing infrastructure could no
longer handle the wastewater created from the rapidly expanding population of the metropolitan region, Chicago went back to the engineering
drawing board.[11]

The Tunnel and Reservoir Project

With the passage of the federal Clean Water Act in 1972, the city of Chicago
could no longer simply send its sewage water downstream in the now
reversed Chicago River, nor could it simply open the gates to allow contaminated waters to flow into Lake Michigan. Under the new federal act,
it would need to develop more effective ways to treat its waste before
it entered the watershed. The proposed response was the Tunnel
and Reservoir Project (TARP) (the deep tunnel). The decision to pursue
TARP was made by a committee comprised of representatives from the
Metropolitan Water Reclamation District.[12]

Under the TARP design, wastewater would flow through local sewers
down interceptor drop shafts into a large tunnel system between 150 feet
and 300 feet below the surface.[13] From these tunnels, the water would
empty into low-lying reservoirs.[14] The overflow water would be stored in
the reservoirs until it could be pumped to wastewater treatment plants
without exceeding those plants' capacity.[15] The underground pipes were
designed to carry the greatest rate of stormwater that had occurred in the
Chicago region in the previous seventy-five years of recorded data.

TARP construction began in 1975 on the tunnels between the towns of
Wilmette and McCook. By 1985, the completed portions of the tunnels
began operation. Today, the system consists of over 109 miles of tunnels
with the largest being about thirty-three feet in diameter. The entire tunnel system of TARP was completed in 2006, and within the next ten
years, three reservoirs with a total holding capacity of over 15.65 billion
gallons will be built.[16]

The TARP project has faced significant scrutiny from its inception. Some
of these concerns have focused on the costs for the project. One critic of
TARP, former United States Senator Charles Percy of Illinois, requested a

formal investigation by the United States General Accountability Office (GAO) in 1980 to review the financial viability of the project. The GAO report documented the spiraling costs of TARP and questioned whether the country could afford such an expensive and precedent-setting public works project.[17] The costs had escalated from $8.5 billion in 1979 to $10.2 billion in 1980, a 19 percent increase.[18] The report concluded that the tunnel portion would likely not be completed based on a weakening economy beset by high inflation, cuts in funding from the United States

Figure 4.2
Inside a TARP tunnel, January 1985. Photograph provided by the Metropolitan Water Reclamation District of Greater Chicago.

Environmental Protection Agency (EPA) for construction grants, and the State of Illinois's funding policy for the program.[19]

This tremendous cost of TARP raises the question of whether there were or perhaps are more cost-effective ways to deal with the problem of contaminated water making its way to the Chicago River and Lake Michigan. One potential alternative would be to reduce the volume of stormwater that enters the combined system by requiring more pervious, nonpaved surfaces and greater water capture and water recycling on site. If there was less stormwater making its way into the combined sewer system, then there would be less need for the storage that TARP provides. The case can be made that the effective implementation of such alternative strategies would in fact get more directly at the true cause of Chicago's combined sewer overflow problem than TARP (which merely deals with the results of these underlying causes) would.

Finally, there were and remain concerns about the environmental impact of TARP. First, as Libby Hill notes, TARP "had and still has its vocal opponents, who not only object to the cost but are also concerned about possible contamination of underground water in the dolomite, or the reverse, the depletion of the groundwater supply by the tunnels."[20] Second, as Libby Hill reports, there are also noxious plumes associated with TARP's operation: "Other evidence that TARP is 'on-line' is the columns of vapor you see (and sometimes smell) at various points along the river. These plumes occur from the warm moist air rising out of the vents from the drop shafts to the deep tunnel. Because the tunnels are not completely drained, residual solids decompose and the musty odor rises with the vapor plumes."[21] TARP's underground storage of untreated, contaminated water therefore has potential adverse impacts on both groundwater resources and airborne conditions.

The Chicago River Revival

In recent years, several factors—including TARP and the closing of riverside industries and stockyards—have contributed to an improvement in the water quality of the Chicago River. This improvement has played a key role in the "river revival" that has taken place in Chicago in recent years.

Throughout the Chicago River ecosystem, fish and wildlife are on the rebound. In the 1970s, fish surveys in the North Branch of the Chicago

River typically found only ten fish species. More recently, surveys have identified over sixty fish species in the river including large-mouth bass, yellow perch, and bluegill, along with a number of nonnative species.[22] Even the Iowa darter, a small fish that has not resided in the Chicago River since the turn of the nineteenth century has returned.

As a testament to how much the situation has improved on the water-quality front, Chicago hosted the Bass Masters tournament, the super bowl of freshwater fishing, in July 2000.[23] Anglers accustomed to the natural splendor of the Everglades and backwaters of the Carolinas found themselves fishing in the shadows of Chicago's skyscrapers.

Other species are also returning. Frogs, crayfish, and turtles can be spotted along the river, and felled trees are evidence of a growing beaver population. Possums, raccoons, and squirrels are also making greater use of the cleaner habitat. Additionally, the river and lakeshore are seeing increasing numbers of birds that pass through the area every year as part of a Midwest flyway.

Additional habitat improvements may make the river even more attractive to animal life. Recent examples of cross-governmental cooperative projects between the Chicago Department of Environment and the United States Army Corps of Engineers include riverbank regrading and plantings of native vegetation at Northside College Prep and Von Steuben High School. The city is installing "fish lunkers" (constructed of trees and rocks that provide underwater habitats) and planting vegetation along the river's edge.

With the improvements in water quality, Chicagoans' view of the river is transforming as well.[24] As the urban population redefines its relationship with the river, recreational use and river access have increased dramatically. A new company, Chicago River Canoe & Kayak, operates a fleet of canoes and kayaks for rent. The city's desire to create additional recreational boat access and paddling opportunities is another indicator the public's attitude toward the Chicago River is changing.

There is now new riverside residential housing along the branches of the river. Even the once highly toxic area known as "bubbly creek" (discussed above) is increasingly lined by new luxury residential properties and riverfront trails.[25] The Ping Tom Park project in Chinatown is another example of how Chicagoans are starting to value the river. Here, the city has converted a twelve-acre former railway site into an accessible riverside park.

In 2003, the city produced a report called *Chicago's Water Agenda*, providing that the river should be managed "for future generations, protected and improved, and managed so that water can continue to sustain us, connect as neighbors, and define our community's role nationally and internationally."[26] The Chicago River is therefore being looked to as a mainstay and catalyst for addressing other urban issues.

Asian Carp and the Chicago River

One consequence of the improved water quality is that the Chicago River now offers a more hospitable environment to invasive species. The Chicago River's polluted water before TARP actually acted as a defense against nonnative species that could enter the river system from the Mississippi River watershed via the Ship Canal.[27] The Chicago River's condition as an ecological deadzone meant that few if any fish could live in the waters between Lake Michigan and the eastern end of the Ship Canal. Hence, along with the return of the native bass fishery, there has been a surge in invasive fisheries—the Asian carp, in particular.

The Asian carp have huge appetites and rapid rates of reproduction, facts that alarm federal and local agencies with jurisdictional authority over the protection of the Great Lakes.[28] The grass carp, native to eastern Asia, was first imported in the early 1960s to aquaculture facilities in Alabama and Arkansas. A decade later, catfish farmers imported bighead and silver carp from Asia, hoping that these species would feed on algae and suspended matter from their ponds.[29] A series of floods in the 1990s resulted in the overflowing of catfish ponds, thereby releasing Asian carp into the Mississippi River basin.[30]

The carp have steadily moved upriver, fast becoming the most abundant species in some areas of the Mississippi River, where they outcompete native fish and decimate fisheries.[31] One unique danger posed by the Asian carp is the fish's tendency to leap out of the water at the sound of loud noises like boat engines.[32] Amazingly, in June 2004 a woman water-skiing near Peoria, Illinois, was struck in the head and rendered unconscious by a massive jumping carp.[33] The Peoria water-skiing incident is indicative of the trend reported in the May/June 2005 edition of *River*

Crossings magazine: "Reports of large jumping silver carp seriously injuring boaters, their equipment, and water-skiers are becoming more frequent.... Recreational anglers and personal watercrafters report a growing number of injuries including cuts from fins, black eyes, broken bones, back injuries and concussions."[34]

The Asian carp have now migrated north to the Illinois River, and as of March 2006, the fish are concentrated near Peoria, Illinois. These adaptive fish are well suited to the temperate waters of the Chicago River and the Great Lakes.[35] Because Asian carp consume a high percentage of their body weight each day and can grow to 100 pounds, the species severely affect the $4.5 billion sport fishing industry of Lake Michigan by outcompeting indigenous fish populations.[36] This reality prompted the Great Lakes Commission to issue a resolution on October 15, 2002, to urge the Army Corps to construct a fish barrier to protect the Chicago River and the Great Lakes ecosystem from Asian carp.[37]

The federal Nonindigenous Aquatic Nuisance Prevention and Control Act of 1990,[38] with amendments contained within the federal National Invasive Species Act of 1996,[39] sets forth national guidelines to combat invasive species.[40] In response to a marked increase in number and types of invasive species inflicting damage on native habitats, President Clinton issued Executive Order 13112 in 1999, which established the National Invasive Species Council with representatives from twelve departments and agencies.[41]

Efforts to stop Asian carp from reaching the Chicago River and Great Lakes have been handled by several federal agencies, including the United States Fish and Wildlife Service, EPA, and the Army Corps working in conjunction with the state of Illinois and the City of Chicago. Other groups such as the International Joint Commission (between the United States and Canada) and the Great Lakes Fishery Commission have joined in efforts to prevent further incursions by the Asian carp.

A Dispersal Barrier Advisory Panel, assembled by the Army Corps, was convened to provide strategies to deal with the various invasive species migrating toward the Chicago River and Lake Michigan via the Ship Canal. Representatives of the Panel included federal bodies such as EPA, the Fish and Wildlife Service, the United States Coast Guard, and the United States Geologic Survey. State agencies included the Illinois Environmental Protection Agency, Illinois Pollution Control Board, Illinois-Indiana Sea Grant

Figure 4.3
Informational poster warning about invasive carp. Courtesy of the Illinois-Indiana Sea Grant program.

program, and Illinois River Carriers Association. The private sector and nonprofits participated as well, including Commonwealth Edison, Smith-Root, Inc., Dosage County Forest Preserve, the Northwestern Illinois Planning Commission, the Great Lakes Fishery Commission, the Great Lakes Fishing Council, the Friends of the Chicago River, and the Canal Corridor Association.

The Panel met several times beginning in 1995 to discuss various options to protect against invasive species. Panel members concluded that chemical, biological, or habitat alteration would not be effective or acceptable due to adverse impacts to the general ecosystem.[42] Members concluded that the best available technology was an electric-shock fish barrier to repel Asian carp trying to the Lake Michigan watershed. An environmental assessment on the viability and impact of such an engineered solution found that an electronic fish barrier would not negatively affect water quality or harm the native fish species that inhabit the canal. Therefore, a finding of "no significant impact" was issued for the electric barrier on December 28, 1999, and the decision to construct a trial barrier was made.[43]

A joint federal and state committee oversaw the construction of the electrical fish barrier near Romeoville, Illinois.[44] The temporary electronic barrier was activated in April 2002 at a total cost of $2.2 million[45]

and for now serves as the Chicago River's primary defense against invasive fish species until the construction of a two-part permanent barrier is
completed.[46]

So far, the temporary barrier appears to be working fairly well. The
Illinois Natural History Survey has tagged and tracked approximately
100 native carp near the Ship Canal.[47] To date, fishery researchers have
netted only one carp north of the barrier. In October 2004, construction
began on the permanent electronic fish barrier.[48] The permanent barrier is
designed to correct the weaknesses identified in the temporary barrier.[49]
This barrier is comprised of two rows of electrodes that stretch cross the
canal approximately 220 feet apart.[50] The new design has a stronger electric field with eighty-three electrodes in each barrier versus the twelve
electrodes in the demonstration barrier.[51] The second row of electrodes
provides a backup system in case of failure of one barrier.[52]

The estimated cost of new barrier is $9.1 million. Congress has author-
ized $6.8 million in funds and the State of Illinois has allocated $1.7 million to the project.[53] The Great Lakes governors have supplied funding
for the remaining nonfederal share of $575,000.[54] The Army Corps will
manage the project.[55] The first part of the permanent barrier goes into
operation in the spring and summer of 2006.

There remain concerns, however, associated with the electric fish barrier. The first concern relates to safety issues when boats come into contact with electrified barriers. In the spring of 2004, a metal cable being
used to tow a barge caused electrical sparks when it passed over the temporary barrier. The shipping industry expressed safety concerns that similar sparks could ignite flammable cargo or cause direct harm to boat
crew that might suffer electrocution.

In response to these safety concerns, the Army Corps and Coast Guard
conducted field tests on the likelihood of this electrical sparking ever hap-
pening again. These results confirmed there was indeed a significant risk
of high-voltage arcing caused by the electrical discharges transmitted to
boats crossing the barrier.[56] These findings led the Coast Guard to adopt
new guidelines for managing boat traffic near the barrier.[57] The Coast
Guard guidelines provide that commercial towboats must use wire ropes,
not metal cables, to ensure electrical connectivity between a barge and
towboat. The guidelines also prohibit loitering of boats or mooring on
the canal bank near the electric barrier and require vessels not to pass,

meet, or overtake in the area. It remains to be seen whether these measures will prevent the reoccurrence of electrocution incidents like the one that occurred in the spring of 2004.

The second concern relates to whether, in the long run, the electric fish barriers are the most ecologically sound approach. Others have suggested that rather than operating an electrified barrier to keep the Asian carp from entering the Lake Michigan watershed, a better-lasting solution would be to separate the waters that are now comingled or connected due to the Ship Canal. More specifically, in May 2003, an Aquatic Invasive Species Summit, attended by more than seventy experts, was held in Chicago to discuss possible responses to invasive aquatic species in the Great Lakes Basin.

In March 2005, Chicago Mayor Richard M. Daley and Fish and Wildlife Service Regional Director Robyn Thorson released an executive summary of the action items that came out of the May 2003 summit. As reported in *River Crossings*:

Recognizing that the impact of invasive species on ecosystems can be permanent and irreversible, the goal of the summit was to find a long-term solution. . . .

The following three actions items were developed:

1. **Separate the Two Basins**—A project should be established that would result in the hydrologic separation of the Great Lakes and Mississippi river basins within 10 years. This long-term solution should consider options including lock modifications and the placement of physical barriers at one or more locations in the Chicago Canal, or other means. Careful assessment is needed in pursuing this approach as navigation, wastewater and stormwater challenges exist.[58]

In effect, this recommendation suggests that further consideration be given to undoing what the Chicago Sanitary District did a hundred years ago when it opened the Ship Canal and reversed the flow of the Chicago River. This recommendation indicates that, with the more complete understanding we now have of the ecological impacts (and economic costs resulting from these impacts) associated with hydrologically linking the Mississippi River and Great Lakes basins, questions are now being raised about whether the maintenance of the current engineered arrangement can still be justified. Perhaps a day is coming when Chicago will have addressed its wastewater and stormwater problems to the degree that serious consideration can be given to reversing (in fact, restoring) the Chicago River to its natural course.

Conclusion

Chicago's legacy of engineered technologies along the Chicago River has helped address the city's water-quality problem. And this improved water quality has led, in turn, to a revival in interest in the river—both as an environmental and recreation amenity and as a location for potential business and residential development. However, Chicago achieved these results in part by externalizing its wastewater problem out of the Great Lakes basin and into the Mississippi River basin. Moreover, the engineering designed to allow this externalization—the Ship Canal—is now serving as a hydrologic pathway for invasive species that may pose an ecological and economic threat equal in magnitude to that formerly posed by pollution concerns.

As such, the question of whether engineered solutions are part of the problem or part of the solution to degraded urban waterways does not point to a simple answer. It depends on what sources of degradation are being considered, on what the timeframe is for evaluating consequences, and on whether the goal of urban river restoration is the narrower objective of reducing toxic dischargers into water or the broader ecological objective of returning waterways to their natural condition.

Notes

1. *The Unofficial Paddling Guide to the Chicago River* 3 (Naomi Cohn ed., 1996).

2. Libby Hill, *The Chicago River: A Natural and Unnatural History* 100–01 (2000).

3. *Id.*

4. *Id.*

5. Steve Jones and John Waller, *Down the Drain: Typhoid Fever City* (2004), *available at* Chicago Public Libraries Digital Collections, http://www .chipublib.org/digital/sewers/history3.htm (last visited Mar. 22, 2005).

6. *Id.*

7. Act of 1889, 1889 I 11. Laws 125. In 1955, the Metropolitan Sanitary District of Greater Chicago changed its name to the Metropolitan Water Reclamation District of Greater Chicago.

8. Hill, *supra* note 2, at 135, *quoting from* Harlow B. Mills, *Man's Effect on the Fish and Wildlife of the Illinois River*, 57 Illinois Natural History Survey Biological Notes 3 (1966).

9. P. Kay Whitlock, *DuPage County's Experience in Storm Water Management, in Water and the City: The Next Century* 361–66 (Howard Rosen & Ann Durkin Keating eds., 1991).

10. Hill, *supra* note 2, at 222–23.

11. Martin Reuss, *The Management of Stormwater Systems: Institutional Responses in Historical Perspective, in* Rosen & Keating, *supra* note 9, at 319–38.

12. *Id.*

13. *Id.*

14. *Id.*

15. William A. Macaitis, *Regional Stormwater Management Trends, in* Rosen & Keating, *supra* note 9, at 306–08.

16. Metropolitan Water Reclamation District of Greater Chicago, *TARP Report Status* (Dec. 1, 2003).

17. Comptroller General of the United States, Letter to Senator Charles Percy from U.S. General Accounting Office, B-201801 (Jan. 21, 1981).

18. *Id.*

19. U.S. General Accounting Office, *Chicago's Tunnel and Reservoir Plan: Costs Continue to Rise and Completion of Phase I Is Unlikely*, CED-81-51 (Jan. 21, 1981).

20. Hill, *supra* note 2, at 223.

21. *Id.* at 225.

22. *Available at* http://www.chicagoriver.org (lasted visited Aug. 5, 2005).

23. Chicago Bass Masters 2000 tournament, *available at* www.lib.niu.edu/ipo/2000/oi000705.html (lasted visited Aug. 5, 2005).

24. David Solzman, *Re-Imagining the Chicago River*, 100 J. Geography 118–23 (2001).

25. Michael Hawthorne, *A Whiff of Success: Million-Dollar Homes along a Long-Polluted Stretch of the Chicago River Fuels New Interest in Cleaning Up Bubbly Creek*, Chicago Tribune, Nov. 21, 2004 (Metro), at 1, 5. Bubbles from decaying animal carcasses discarded from slaughterhouses cause this section of the Chicago River to bubble up. Some of these odorous smells may be a result of untreated sewage. *Id.*

26. Chicago's Water Agenda 2 (2003), *available at* http://egov.cityofchicago .org/city/webportal/portalDeptCategoryAction.do?BV_SessionID=@@@@12341 43437.1163206785@@@@&BV_EngineID=ccccaddjgdehggmcefecelldffhdfgn .0&deptCategoryOID=-536890236&contentType=COC_EDITORIAL&top ChannelName=SubAgency&entityName=Conserve+Chicago+Together&dept MainCategory OID=-536889943.

27. Jerry L. Rasmussen, *The Cal-Sag and Chicago Sanitary and Ship Canal: A Perspective on the Spread and Control of Selected Aquatic Nuisance Fish Species*, United States Fish & Wildlife Service, Jan. 1, 2002, at 1–3.

28. Reports included a 600-fold increase in carp numbers between 1999 and 2000 in LaGrange, Illinois, in the Illinois River navigation dam, which links to the Chicago River. Mississippi Interstate Cooperative Resource Association, *Chicago Summit Generates Possible Solutions to Invasive Species Issues*, River Crossings, 11, May/June 2005, at 1–3; Rasmussen, *supra* note 27, at 1–3; *see also Asian Carp Invasion of the Upper Mississippi River System*, Upper Midwest Environmental Sciences Center Project Status Reports (2000–2005).

29. U.S. Environmental Protection Agency, *Great Lakes, available at* http://www.epa.gov/glnpo/invasive/asiancarp/index.html (last visited Oct. 15, 2004).

30. *Id.*

31. *See supra* note 28.

32. Dan Wilcox, St. Paul District, U.S. Army Corp of Engineers, *Invading Asian Carp Post Unusual Threat*, Engineer Report Update (May 2004), *available at* www.hq.usace.army.mil/cepa/pubs/may04/story9.htm (last visited Dec. 15, 2004).

33. Tom Meersman, *Jumping Carp Maul Boaters on Illinois River in Peoria*, Minneapolis–St. Paul Star Tribune, June 18, 2004, *available at* http://casper startribure.net/articles/2004/06/20/news/national/701a2b874ad0c5b287256eb6 006c8d64.txt (last visited Dec. 2005). According to the victim, " 'I'm sitting there and all of a sudden this big fish flops out of the river literally and hits me right between the eyes. I'm not kidding. It knocked me completely out.' [The victim] . . . revived quickly, but she found herself floating face down in the river, bleeding profusely. She saw her watercraft floating away in the current, heading toward a towboat that was blasting its horns. She passed out again, but a nearby boater, alerted by the warning blasts, came to her rescue. [The victim] suffered a broken nose, concussion, black eye, injured back and a broken foot. Other boaters along the Illinois, Missouri and Mississippi rivers have reported dislocated jaws, facial cuts, broken ribs and serious bruises. Hundreds have been startled as the thin-skinned carp shot into their boats and flew to pieces as they hit seats, coolers, fishing equipment and depth finders."

34. *Asian Carp Risk Assessment Completed*, River Crossings, May/June 2005, at 4.

35. Mississippi Interstate Cooperative Resource Association, *supra* note 28.

36. Rasmussen, *supra* note 27.

37. Great Lakes Commission, *available at* http://www.glc.org/about/resolutions/02/asiancarp.html (last visited Mar. 2, 2005).

38. Pub. L. No. 101–646, tit. I, 104 Stat. 4761, 16 U.S.C. 4701 (enacted Nov. 29, 1990). It established a new federal program to prevent and control introduction of invasive species.

39. Pub. L. No. 104-332, 110 Stat. 4080, 16 U.S.C. 4701–4751 (enacted Oct. 26, 1996).

40. U.S. Environmental Protection Agency, *available at* http://www.epa.gov/grtlakes/glwa/usreport/part5.html (last visited Dec. 15, 2004).

41. Exec. Order No. 13,112, 64 Fed. Reg. 6183 (Feb. 3, 1999).

42. Army Corps of Engineers, *Aquatic Nuisance Species Dispersal Barrier Demonstration Project: Chicago Sanitary and Ship Canal between Lemont and Romeoville, Cook and Will Counties,* Illinois Environmental Assessment (Aug. 1999).

43. Army Corp of Engineers, *Finding of No Significant Impact for the Aquatic Nuisance Species Barrier Demonstration Project: Chicago Sanitary and Ship Canal between Lemont and Romeoville, Cook and Will Counties,* Illinois Environmental Assessment (Dec. 28, 1999).

44. U.S. Environmental Protection Agency, *Great Lakes, available at* http://www.epa.gov/glnpo/invasive/asiancarp/index.html (last visited Oct. 15, 2004).

45. *Id.*

46. A demonstration barrier was necessary to ensure that different sizes and species of fish would react differently to different electric fields. Thus, the electric charge had to be carefully calibrated to stop movements of the carp. Mississippi Interstate Cooperative, *supra* note 28.

47. Environmental Protection Agency, *Great Lakes: Big-Head Carp, Asian Carp and the Great Lakes, available at* http://www.epa.gov/glnpo/invasive/asiancarp/index.html (last visited Mar. 2, 2005).

48. *Id.*

49. *Id.*

50. *Id.*; *see also* Army Corps of Engineers, *supra* note 42.

51. Army Corps of Engineers, *supra* note 42.

52. Environmental Protection Agency, *supra* note 47.

53. Environmental Protection Agency, *Federal Funding Available for Enhanced Protection Against Asian Carp,* Press Release (Oct. 13, 2004).

54. *Id.*

55. The operating cost of running a high-voltage wire for multiple years will be significant.

56. Department of Homeland Security, Coast Guard, Regulated Navigation Area, Chicago Sanitary and Ship Canal, Romeoville, IL, Final Rule, 33 C.F.R. pt. 165 FR Doc. 05-24538 (Dec. 27, 2005).

57. *Id.*

58. *Asian Carp Risk Assessment Completed, supra* note 34, at 45.

5

Bankside Salt Lake City

Ron Love

City Creek flows out of City Creek Canyon into the northern edge of the central business district in downtown Salt Lake City, Utah. City Creek Canyon is contained in low-lying foothills of the Wasatch Mountain Range, which run along the east side and north end of the Salt Lake Valley.[1] The canyon rises from the valley floor, which sits at 4,200 feet,[2] to the head of the canyon (at 9,400 feet) and extends twelve miles into the foothills of the Wasatch Mountains.[3] City Creek drains 19.2 square miles of watershed.[4]

City Creek Canyon varies from rock outcroppings (where alluvial deposits are readily seen) to lush areas of dense trees (deciduous trees at the lower elevations and conifers at the canyon head).[5] Shrubs (mostly of scrub oak) and grasses (which cover vast areas of the gently sloping canyon) are prevalent along the middle area of the canyon.[6] The mouth of City Creek Canyon, with its tall and stately cottonwood trees, has been part of Salt Lake City since the city was incorporated.[7] The bulk of lands located in the creek and the canyon have been annexed in modern times to allow Salt Lake City Public Utilities Department strong management jurisdiction over the watershed.[8]

This chapter addresses the background of this historically significant creek, its encapsulation early in the twentieth century, and a modern attempt to daylight the creek using legislation originally enacted as part of the Clean Water Act. It begins by tracing the background leading to the national movement to restore rivers and streams, which began in the 1970s and continues to gain momentum. It looks briefly at the Brownfields Showcase Project, which spurred the daylighting, and then explores in detail the United States Army Corps of Engineers' efforts under the ecosystem restoration program toward daylighting the creek, as well the Urban Rivers Restoration Initiative, a congressionally directed joint initiative of the Army Corps and the United States Environmental Protection Agency (EPA).

Preliminary efforts have been undertaken on a project to unentomb City Creek. The Daylighting Project is explored in detail below to provide insights to others who are involved in urban stream-restoration efforts by examining some of the lessons learned over the ten-year history of this project. The project's experience with City Creek suggests the pitfalls of relying too heavily on federal funding and too little on local sources of support to drive urban waterway-restoration projects.

History and Encapsulation of City Creek

Both City Creek and City Creek Canyon have played mayor roles in the cultural and economic development of Salt Lake City. The early Mormon pioneers, led by Brigham Young, camped at the mouth of the canyon when they entered the Salt Lake Valley,[9] which was part of Mexico at that time.[10] Young's first house and farm were on land that was deeded to him by the territorial legislature in 1857.[11] The pioneers used City Creek as a source of drinking and irrigation water.[12] They also used it to power a sawmill, a flour mill, and a silk plant. All such activities contributed to the establishment of Salt Lake City as the territorial and later state capital.[13]

The high quality of City Creek's untreated water also played a major role in the economic development of Salt Lake City from the pioneer days to the middle of the twentieth century, when the first water-treatment plant was established five miles up City Creek Canyon.[14] Early in the development of the territory, the overarching role of water became clear. In 1995, Thora Watson noted in her book *The Stream That Built a City*: "Control over water has ultimately become tantamount to controlling the destiny of the land and the people who settle on the land."[15]

The excellent drinking water from City Creek was sold outright or traded for irrigation water from the nearby Jordan River, which was degraded initially by the various farming and agricultural activities that quickly sprung up along this central-valley river and which was subsequently contaminated by the mining efforts in the foothills and mountains along both sides of the valley.[16] City Creek water thus became the standard in the valley for quality potable water.

Until the end of the first decade of the twentieth century, City Creek was bifurcated at the mouth of City Creek Canyon at the north end of Salt Lake City. Both prongs of the creek eventually reached a confluence with the Jordan River, two miles west of the mouth of the City Creek Canyon.[17] One leg of the creek flowed due west to the Jordan River along what is now North Temple Street; the other leg flowed south for a half mile before making a right-angle turn and flowing to the Jordan River along what has become the Fourth South Street right-of-way.[18] This latter branch of the stream provided irrigation to crops grown on what is now Washington Square, a ten-acre park where City Hall was constructed between 1891 and 1893.[19]

The volume of City Creek water is subject to seasonal climatic conditions. According to the 1999 *Salt Lake City Watershed Management Plan,* "Characteristically, there is a gradual rise in flows throughout April with a marked increase early in May as temperatures increase. Flows decrease through June and July, and stabilize during August. The moderate fluctuations of the Creek are attributed to the nearly constant sun exposure to snow pack on the gentle slopes, and the cavernous nature of the subsurface limestone from which the canyon's springs rise. The [historic] average annual yield for the creek is 11,749 acre feet."[20]

Pictures available from the Utah State Historic Preservation Office show the effects of the fluctuations of City Creek during the first decade of the twentieth century.[21] During the peak flows in the spring, flooding of the streets and buildings on streets without sidewalks, curbs, or gutters was common along both branches of City Creek below the mouth of the canyon.[22] As the water drained from the streets, it carried with it animal waste and harmful bacteria.[23] During the dry periods of late summer, mosquito problems appeared when the water formed stagnant pools, which created breeding grounds for these and other pests.[24]

In 1909, Salt Lake City entombed City Creek in an underground culvert that ran from the mouth of City Creek Canyon to the Jordan River underneath North Temple Street.[25] *The City Creek Master Plan,* completed by the Salt Lake City Planning Division in 1986, points out that this was done to "protect the water supply and prevent accidental drowning."[26] Public health concerns therefore were the key rational for

Figure 5.1
Encapsulation and entombment of City Creek, circa 1909. Photograph from the archives of the United States Army Corps of Engineers.

burying the stream. Thomas Alexander in his recent book, *Grace and Grandure: A History of Salt Lake City,* informs us that in the following decade, the City Council designated City Creek Canyon as a "nature park" to preserve the quality of City Creek water and keep the watershed free of industrial activities.[27]

Development of City Creek Restoration Initiatives

It is difficult to say exactly when the current movement to preserve our nation's rivers and streams and restore aquatic, riparian ecosystems began. The environmental movement that grew from the activism of the 1960s led to the establishment of the EPA and the passage of a string of laws that armed citizens and communities with the means necessary to prevent further degradation of rivers and streams throughout the United States. Key among these laws was the National Environmental Policy Act (NEPA) enacted by Congress in 1969 and the Federal Water Pollution Control Act enacted in 1972, which became commonly known as the Clean Water Act.[28]

In his book *Endangered Rivers and the Conservation Movement,* Tim Palmer explained the movement to restore our waters in this way: "A revolution in attitudes about rivers moved through the country and touched every stream. The late 1960s and early 1970s brought powerful ingredients for change: a growing sense of scarcity, the environmental movement, activism by conservationists and landowners, application of science and economics coupled with publicity, recreation use, and tight money—all contributing to a national movement to save threatened rivers."[29]

The decade following the enactment of NEPA saw a boom in outdoor recreation throughout the nation. Boaters of nonmotorized craft took the rivers and streams, and outfitters cropped up in every state, particularly near whitewater stretches of major rivers.[30] In Moab, Utah, for example, a number of current outfitters' advertisements indicate that they were formed in the 1970s.[31] According to Palmer, "Once they experienced a wild river—any wild river—people could understand conservationists wanting to save a similar place.[32] Paddlers and river guides became activists, and like the hikers and climbers of the 1950s who matured into the wilderness preservationists of the 1960s, these river runners forced their way into the political process."[33]

Citizen involvement throughout the history of the United States increases as people become convinced that a resource that they hold dear is threatened. The movement to protect wild rivers for recreational uses came about gradually as grassroot activists were able to gain the attention of members of Congress through volunteer campaigns aimed at the latest environmental threat, lawsuits attacking elements of environmental impact statements, and public campaigns typically motivated by a perception that the last vestiges of a particular resource were gravely threatened. Growth in the western United States became the pivotal issue for those involved in saving rivers and streams in general and led specifically to emphasis on urban riparian issues.

Robert Gottlieb, a dedicated student of water policy in the western states (and one of the other contributing essayists to *Rivertown*), explained in his book *A Life of Its Own: The Politics and Power of Water* that in the 1980s and 1990s, "Slow growth positions were increasingly adopted by local community groups operating either on the edges of or separate from the mainstream environmental organizations. The concerns they raised were primarily urban ones, such as congestion, pollution, the lack of green space and the deterioration of everyday life. These movements focused on how their neighborhoods and communities were affected by housing and transportation, toxic dumps, and air and water quality, issues not ordinarily found within the traditional environmental agenda."[34]

In 1986, Tim Palmer acknowledged that "There is a new attitude about rivers, but to make the change permanent will require the energy of all those affected by the loss of special river places: park and wilderness enthusiasts, rangers, landowners, naturalists and ecologists, river runners, people who just like rivers, and those who find metaphysical and spiritual power in the free-flowing water."[35] Fortunately, this attitude—enhanced by sustainable-growth issues—was embraced and fostered in federal legislation that led directly to the proposal to daylight City Creek in Salt Lake City.[36]

Two key pieces of federal legislation and a joint initiative by the EPA and the Army Corps have been employed in the effort to daylight City Creek. As explained below, the Clean Water Act, the Brownfields Redevelopment Act and the Urban Rivers Restoration Initiative each played a key role in this undertaking.

The Gateway Brownfields Pilot Project

The Brownfields Program[37] was established in 1995 as an administrative program of the EPA. It received the blessing of legislative action when Congress passed the Brownfields Revitalization Act of 2001,[38] which allowed the establishment of the Brownfields Cleanup and Redevelopment Initiative.[39] A mostly abandoned, 650-acre railroad yard on the western edge of the Central Business District in downtown Salt Lake City was selected as a pilot project (Brownfields Pilot). The concept of revitalizing the area located in an historic district of Salt Lake City with more than 100 years of recorded industrial use had been bantered about by the Salt Lake City Planning Division since the early 1980s and started appearing in planning documents in the mid-1980s.[40] To explore redevelopment options, Salt Lake City contracted for a Gateway Visualization Plan in 1993.[41] The plan was used to determine redevelopment plans for the area.[42]

The designation of the Gateway Area as a Brownfields Pilot accelerated the redevelopment of the region with a series of grants totaling $900,000 and, perhaps most important, the loan to Salt Lake City of an environmental scientist from EPA Region 8 in Denver (EPA Denver), Stephanie Wallace.[43] On her three-year assignment to the Salt Lake City Redevelopment Agency, funded by the EPA, Wallace provided oversight to the environmental cleanup of the Brownfields Pilot area and coordinated a request with the Army Corps to daylight City Creek.[44]

One-third of the proposed daylighting occurs in the designated Gateway Brownfield Showcase Pilot area.[45] This allowed the use of federal Brownfields grant money to fund a portion of the environmental and hydrological studies necessary for the completion of a reconnaissance study by the Army Corps.[46] This fortuitous tie to Utah's Gateway Visualization Plan brought the daylighting of City Creek to the forefront of public attention by its inclusion in information and presentations made to national audiences.[47]

The City Creek River Ecosystem Restoration Effort

The Proposed Daylighting Project

The Daylighting Project, as it has evolved to date, calls for diverting water from the existing culvert to the point of daylighting just west of a recently completed $375 million Brownfields redevelopment project consisting of an outdoor commercial section with ninety shops and restaurants, twelve stories of condominiums, a planetarium, a convention space, fountains, twelve movie theaters, and planning for 10,000 new residential units.[48] The daylighted creek would then meander approximately 7,900 linear feet in a dirt-lined open channel approximately three feet deep, ten feet wide at the top, and two feet wide at the bottom.[49] The proposed channel slope would be 2.7 feet horizontal to one foot vertical. The riparian area containing the creek would consist of approximately thirteen acres in a strip that varies from eighty feet to 150 feet wide as it passes under a major freeway and through an area of the city now in transition.[50] Zoning in the area is industrial, commercial, and residential.[51]

The Daylighting Project corridor will be revegetated with native, riparian, and upland species.[52] The Daylighting Project alignment will follow an existing railroad track that is to be realigned approximately 330 feet to the north, adjacent to an existing main line track.[53] A ten-foot wide paved, pedestrian, and bike trail will also traverse the area and serve as a maintenance road.[54] This trail would connect downtown Salt Lake City with the Jordan River Parkway, a regional trail that connects to the Bonneville Shoreline Trail, another recipient of Brownfields restoration funds, which, when completed, will run the full length of the Salt Lake Valley—from Brigham City to Provo, Utah, a distance of over 100 miles.[55] A planned two-mile, rails-to-trails project will connect the Jordan River trail to yet another regional trail, the Pratt Trail, which is constructed on an abandoned railroad track running from the Salt Lake Valley all the way to Evanston, Wyoming, a distance of 100 miles.[56] The trail along the proposed daylighted stream will provide alternative means of transportation between various far-reaching areas of the valley and the downtown areas of the Salt Lake City and, according to a fact sheet published by the Army Corps South Pacific Division, "would allow ecosystem appreciation opportunities to site visitors."[57]

The major objective of the Daylighting Project is the restoration of approximately thirteen acres of riparian habitat, with a focus on the creation of emergent, riparian, and upland grasses and trees.[58] Local native species will be used for the revegetation efforts to provide appropriate habitat and foraging opportunities for wildlife, insects, and fish.[59] Salt Lake City participated in the development of planning objectives for the Daylighting Project, establishing the framework for the aquatic restoration study:

· Reestablish a portion of the City Creek aquatic and riparian habitat—lost a century ago—between the regionally significant Jordan River and downtown Salt Lake City;
· Create aquatic and terrestrial habitat with associated wildlife values to target feeding and cover for migratory birds. The habitat will potentially include emergent marsh, riparian forest, and upland native grasslands. Plants will provide a diverse structure to provide foraging opportunities for various guilds of birds, such as fruits and seeds, insect and fish. Landscaping will provide shade in order to moderate water temperatures;
· Landscape with low water use plants, using native species where feasible to serve as a model for low water use;
· Provide environmental education and stewardship opportunities in Salt Lake City;
· Increase vegetative open space to improve aesthetic values for nearby residents and businesses and preview a connection to regional green space areas;
· Reestablish the surface connection between City Creek and the Jordan River;
· Establish multi-use trail/maintenance access;
· Evaluate the potential for an urban fishery;

for more natural projects, but there are benefits to fish, passerine birds, and small mammals like raccoons, opossums, and (yes) skunks that can be quantified. Whether or not the Project is approved will likely be decided by SPD [Army Corps South Pacific Division], which hasn't been very restrictive lately regarding the approval of restoration projects that meet basic policy requirements. There are no clear criteria by which to decide whether a restoration project is justified or not. Nonetheless, everything possible should be considered to maximize habitat and other ecological values within the constraints of the Project setting.[65]

Recommendations from the U.S. Fish and Wildlife Service also pointed to the challenges faced by attempting a restoration in an urban environment:

Because of the urban nature, small scale, and restrictions imposed by a narrow right or way of the project, all the recommendations made here are important in justifying § 206 monies being applied to this endeavor. That is not to say there is no flexibility in the design and implementation of theses recommendations, only that the exclusion of one or more of them may reduce the habitat value of the project more than it would if the project area were larger, more diverse, and less impacted by the surrounding environment.[66]

In an attempt to show the level of support and to respond to criticism of the Daylighting Project, the City of Salt Lake City established a working group of representatives from potentially affected organizations. The working group included Trout Unlimited, Union Pacific Railroad, Utah Heritage Foundation, Western Wildlife Conservancy, Save Our Canyons, City Creek Coalition, Tree Utah, Utah Department of Environmental Quality, Utah Department of Fish and Wildlife, Utah Power and Light, and Questar Corporation.[67] Initially, when the Army Corps representatives conferred on the Daylighting Project, Salt Lake City invited members of the working group to provide input. The attendance of interested parties from the public, corporate, and nonprofit sectors allowed Army Corps representatives to better gauge the degree of local and state support for the Daylighting Project.

This support, however, was not for the restoration of City Creek where it flowed a century ago. The former creek corridor, according to early proposed language for the Project Documentation Review, through which the original stream "formerly flowed now appears as a major urban heat island on thermal imaging maps created for Salt Lake City by NASA in large part due to the almost complete lack of vegetation in the area."[68] The proposed daylighting was therefore not sited along the original creekbed but instead was sited along a railroad right-of-way through an underdeveloped neighborhood.[69]

Negotiations with Union Pacific

Early in the planning process, there were concerns that the high value of urban real estate in the area where the daylighting was proposed would prove a significant hurdle. More specifically, there could be high costs associated with public acquisition of the land where the daylighting was to take place, and these could push the project beyond the $7 million available pursuant to section 206. These concerns were ameliorated somewhat through successful negotiations between Salt Lake City and Union Pacific.

The city and Union Pacific entered into talks for the development of an agreement that would facilitate a land trade that would net the city 11.5 acres of land necessary for the Daylighting Project.[70] In exchange for providing these 11.5 acres to the city, the city agreed to provide assistance (both politically and monetarily) to straighten the Folsom Street line that currently curves its way through the Grant's Tower area. The straightened track will enable faster train speeds and help alleviate bottleneck conditions. All the land necessary for the recreation of City Creek, with the exception of one parcel consisting of less than one-third of an acre, historically belongs to either Salt Lake City or Union Pacific.[71] The agreement represents neither a magnanimous gesture on the part of the railroad nor the absence of a problematic relationship between the City and Union Pacific. The relationship between the two has been tenuous at best.

Salt Lake City unsuccessfully sued to stop Union Pacific from reopening an unassociated, previously abandoned section of rail track through an economically depressed residential neighborhood at approximately the same time that the city and the Army Corps needed to access railroad property to perform environmental, geotechnical, and landscape revegetation soil testing.[72] It was just prior to this litigation that disputes over property ownership between Salt Lake City and Union Pacific came to a head.[73] This dispute resulted in a nearly twelve-month delay in gaining access to the railroad property for the necessary environmental and geotechnical studies in the proposed right-of-way for City Creek. As explained below, this delay unfortunately had significant adverse impacts on the Daylighting Project.

Shortfalls, Turnover, and Cost-Benefit Analysis

Without the twelve-month delay resulting from the access dispute with Union Pacific, the Feasibility Study and Army Corps-City funding agreement would likely have been completed in short order, and the $7 million in section 206 funds would likely have been committed to the project. As it happened, the twelve-month delay coincided with the escalation of United States military involvement in Afghanistan and Iraq and with the decision to reallocate portions of section 206 funds for overseas national security purposes.

All Army Corps section 206 projects without completed feasibility studies were conducted with funds from a continuing resolution passed by Congress in October 2003.[74] In early February 2004, the Army Corps fiscal-year budget appropriations for 2004 were announced. The amount funded for incomplete section 206 projects was woefully insufficient.[75] The Army Corps notified Salt Lake City in early February 2004 that funds for the completion of the Feasibility Study were unavailable.[76] Work on the study came to a halt.[77]

As of this writing, funds for the completion of the Feasibility Study have not been allocated. The study is two-thirds done but requires additional funds to complete.[78] City, state, and local private interests have so far nonetheless been unable or unwilling to step forward with the additional funding needed to complete the work. The Utah congressional delegation representing Salt Lake City has attempted to free up additional section 206 funds to cover the shortfall but so far without success.[79]

The current funding shortfall (for completing the feasibility study and implementing the proposed daylighting project) is not the only threatening factor. Even if section 206 funds are reallocated, the question then arises as to which people at the Army Corps can or will carry the work forward. Historically, the turnover of personnel in the Army Corps Sacramento District has adversely affected the Daylighting Project because new members of the team require time to familiarize themselves with the Daylighting Project. Often, the remaining members of the team have had to rehash previous assumptions, combat reoccurring objections and misunderstandings, and deal with numerous differences of opinions about how best to accomplish the complicated business of ecosystem restoration. Occasionally, the personnel turmoil, which included the change of project managers for the city and the Corps, leads to the incorporation of ideas that strengthen the feasibility of the Daylighting Project. More often than not, it leads to the restatement of a previous rationale.

One salient exception to personnel turnover has been the Army Corps local project engineer, Scott Stoddard, the intermountain representative stationed in Bountiful, Utah. Stoddard's institutional knowledge and commitment to the Daylighting Project have helped provide at least some degree of continuity in the face of significant personnel changes in the Army Corps and also in the various state and federal agencies involved.

Part of the problem is that only a limited number of people at the Army Corps have a solid understanding of both the science and economics of ecosystem restoration. Institutionally, the Army Corps has developed good publications on this topic, such as the engineering pamphlet *Ecosystem Restoration: Supporting Policy Information*. This pamphlet provides the following guidance on ecosystem restoration:

Ecosystem Restoration is a primary mission of the Civil Works program. Civil Works ecosystem restoration initiatives attempt to accomplish a return of natural areas of ecosystems to a close approximation of their conditions prior to disturbance, or to less degraded, more natural conditions. In some instances a return to pre-disturbance conditions may not be feasible. However, partial restoration may be possible, with significant and valuable improvements made to degraded ecological resources. The needs for improving or re-establishing both the structural components and the functions of the natural area should be examined. The goal is to partially or fully reestablish the attributes of a naturalistic, functioning and self-regulating system.[80]

The peer review of the Army Corps Study notes also contains some encouraging information. In his review, Scott Miner, the previous ecosystem restoration specialist in the Planning Division of the Army Corps, Sacramento District, pointed out additional sound reasons for conducting the Daylighting Project:

There is widespread public interest in the restoration of urban creeks for their aesthetic, recreational and educational, as well as ecological, benefits (see www.urbancreeks.org, for example). Successful completion of a high-visibility project like City Creek could provide a great deal of positive publicity for the Corps, leading to future restoration projects far beyond Salt Lake City.[81]

Despite the Army Corps' recognition of the importance of ecosystem restoration, the daylighting proposal is still a relatively new type of project for the Corps. In the past, the Army Corps have been involved primarily in concrete encasement and culverting of streams, not the *de*encasement and *de*culverting proposed with the City Creek project.

Certainly, individual staff persons within the Army Corps are support-ive of this restoration work, but it is an institution in flux. The high turnover of Army Corps staff involved in the City Creek project means that there is a constant need to reeducate incoming Army Corps personnel about the project and that Army Corps enthusiasm for the daylighting efforts varies with these different personnel. These are not ideal conditions for keeping the momentum going forward during this period of budgetary shortfalls.

A final obstacle to maintaining Army Corps support—both at the staff level and in terms of congressional appropriations for section 206—is the way in which cost-benefit analysis is applied to ecosystem restoration projects. Cost-benefit analysis involving qualitative elements such as the quality of habitat can be problematic to Army Corps methodology. The Army Corps guidance addresses that by recognizing its subjective nature and encouraging the development of experience and use of professional judgment in making the analysis.[82] Habitat values are assigned various elements of the ecosystem plan based on their benefit and harmonious fit with other elements as well as their completeness, effectiveness, effi-ciency, and acceptability.[83] Cost effectiveness is also listed by the Army Corps guidance as a major player in this analysis:

An ecosystem restoration plan should represent a cost effective means of address-ing the restoration problem or opportunity. It should be determined that a plan's restoration outputs cannot be produced more cost effectively by another alterna-tive plan. Cost effectiveness analysis is performed to identify least cost plans for producing alternative levels of environmental outputs expressed in non-monetary terms. Incremental cost analysis identifies changes in costs for increasing levels of environmental output. It is used to help assess whether it is worthwhile to incur additional costs in order to gain increased environmental outputs.[84]

These comments acknowledge that nonmonetary ecological outputs merit Army Corps support but do not provide Army Corps staff or ecosystem restoration advocates with much help for making the case for why such ecological outputs make good economic sense. For instance, improved environmental amenities such as restored urban waterways can translate into increased property values both for residences and retail/office investment. These are the types of cost-benefit arguments that would lend more meaningful political support to the case for fund-ing by the Army Corps (as well funding by nonfederal entities and the private-sector interests that may be in a position to help underwrite the daylighting work). This type of analysis, however, is something new for the Army Corps.

Signs of Hope

Although the budgetary shortfalls, personnel turnover, and constrained cost-benefit analysis have proven obstacles to moving the City Creek daylighting project forward, there is also cause for cautious optimism that the work will eventually be able to proceed.

Resolving Toxic Concerns

As noted above, approximately one-third of the Daylighting Project area falls in the Gateway Brownfields Showcase Pilot area. This bodes well for the Daylighting Project. The daylighting of City Creek has been mentioned in various briefings, fact sheets, and poster displays at national and international Brownfields conferences.[85] Scott Stoddard, the Intermountain representative of the Army Corps, presented a briefing on the Daylighting Project at the first National Ecosystem Restoration Conference held in Orlando, Florida, in December 2004.[86] Since it is associated with a Brownfields pilot on showcase project of considerable magnitude, it is featured prominently at Brownfields events.

This association has economic benefits, as well. Brownfields funds were authorized for some of the environmental testing required by the feasibility Study.[87] The Gateway Brownfields area is also one of the Salt Lake City Redevelopment Agency (RDA) designated target areas.[88] The RDA has budgeted nearly $1 million to the Daylighting Project, once it is approved for design.[89]

Soil and groundwater testing was completed under contract with the Army Corps in December 2002.[90] The testing included sampling at seven locations along the proposed 1.5-mile City Creek alignment.[91] The results of this testing confirmed that toxics should not be an obstacle to the daylighting effort.[92]

In terms of soil samples, results showed that, "With the exception of arsenic, metals concentrations were below EPA action levels[93] for soils at industrial sites. The report on the testing, however, concluded that even the arsenic results should not be a concern since such arsenic concentrations are generally naturally occurring."[94]

In terms of groundwater conditions, the toxics analysis indicated that additional action was not warranted.[95]

The Urban Rivers Restoration Initiative

Salt Lake City applied for and received a grant from the federal Urban Rivers Restoration Initiative (URRI) for citizen outreach associated with the creek ecosystem restoration.[96] The URRI is a joint initiative of the EPA and the Army Corps created in the summer of 2002 by the signing of a memorandum of understating (MOU).[97] The URRI was created to facilitate a more deliberate and coordinated effort of the two agencies toward remedial water quality and environmental restoration of urban rivers and streams. The URRI has provided grants for eight pilot projects nationwide.[98]

The URRI has strong synergy with several current major federal initiatives, including the Brownfields redevelopment initiative, the total maximum daily load initiative,[99] the natural-resource damage assessment and restoration program, new ecosystem restoration and protection, and aquatic ecosystem restoration authorities provided to the Army Corps in recent Water Resources Development Acts.[100] Article III of the 2002 MOU between the EPA and the Army Corps for the creation of the URRI defines the scope of the program:

In order to begin an evaluation of this urban rivers cooperative approach it is proposed that eight demonstration pilot projects be announced and undertaken during the next 12 months. The pilot projects will include, but not be limited to, projects for water quality improvement, contaminated sediment removal and remediation and riparian habitat restoration.[101]

All eight of the pilot projects stress partnership formation among various government, not-for profit, and for-profit entities with the goal of mitigating environmental insults and restoring urban waterways. At the National URRI Conference in Salt Lake City in June 2003, it was announced that no new pilot projects would be selected under the 2002 MOU set to expire in June 2004.[102] The joint efforts and resources of EPA and the Army Corps would be focused on existing pilot projects.[103] This decision benefited the Daylighting Project because additional funds were provided to enhance the activities under the URRI Pilot Program.[104]

The 2002 MOU clarifies that nothing in the agreement will change any statutory or regulatory obligations:

This agreement establishes a mechanism of cooperation and coordination, and expresses the intent of the signatory agencies to work together to resolve any conflicts using, as appropriate, consensus building and collaborative decision-making to find common ground and identify practical solutions. Success of this agreement will be evidenced by the efficient accomplishment of each agency's statutory requirements within areas of mutual concern in a timely manner and by minimizing misunderstandings and duplication of effort.[105]

The initial memorandum of understanding was established with an end date of June 2004.[106] At the time of this writing, a new memorandum of understanding, which would allow the extension of the URRI program, had been forwarded for signatures.[107]

Citizen Outreach and the Small Area Master Plan[108]

As a result of the link to the Gateway Brownfields project, there was considerable interest in City Creek daylighting at EPA offices in Denver. This, coupled with the study by the Army Corps, led to an invitation to Salt Lake City to apply for the pilot URRI grant (which as noted above, it received). The following excerpts from the grant application explain how the funds will be used for community outreach:

Salt Lake City intends to use the funds for community outreach and the development of a Small Area Master Plan (SAMP). The SAMP will bring a number of players from local, federal and state agencies, the community, businesses owners, and property owners, and other interested parties together to formulate a plan that will set the tone for the growth and development of the area once the creek is daylighted and allow the community input into the design and planning of the daylighted creek. The planning area consists of approximately 140 acres currently crisscrossed by rail road tracks, with a mixture of residential, industrial and commercial [uses]. The SAMP will provide a series of integrated recommendations for business and residential land uses, recreation uses, and multi-modal transportation needs in an area that has been in transition for many years.

The Daylighting of City Creek will restore an aquatic uses in the area, provide native plant and animal habitat, and provide recreation opportunities that do not now exist in an area that shows severe signs of blight.

Anticipated measures of success: Salt Lake City has embraced the concept of Sustainable Development in its master planning and land use policies. The benefits of this project will be measured in the reduction of crime in the area, by the expansion of riparian habitat, and by the economic benefits including the revitalization of an area in transition.[109]

Using the URRI pilot grant, Salt Lake City was able to direct the efforts of a firm under contract with EPA Denver toward the development of this

SAMP.[110] The initial concept was addressed at an open house held in the target area.[111] The response to the open house in terms of citizens (mainly residents and business owners in the neighborhood) was relatively good. Twenty individuals signed up to participate as working group members or steering committing members.[112] At subsequent meetings, that interest has remained high, and citizen participation has exceeded the expectations of city planners.[113]

The development of the SAMP was infused by the grant of an additional $50,000 under the URRI in 2005. This allowed for the expansion of the SAMP to add a visualization component and a market study to determine additional needs of the community. Completion of the SAMP is scheduled for the summer of 2006, at which time it will be presented to the Salt Lake City Planning Commission for approval and forwarding to the City Council for adoption.

Conclusion: A Search for Local Solutions

After six years of effort by various agencies, community activists, and City Planners on the Daylighting Project, City Creek remains entombed, and its daylighting still only a plan. Moreover, because of the prospect that Army Corps support might be restored, there has so far been an unfortunate reluctance to explore other sources of potential funding.

At this point, Utah state government, local government, and local private interests have not fully committed to the project. When faced with the Army Corps budgetary shortfalls, these other entities and interests have not stepped forward to fund the work so that it can proceed. This does not necessarily imply that they are opposed to the project or that they would not like to see it move forward, but it does reveal that daylighting City Creek has not been a major priority at the local level or presumably the funding to proceed would have been made available regardless of the status of Army Corps section 206 appropriations.

In hindsight, effective congressional intervention by supporters of the City Creek daylighting project employed early on may have helped maintain the section 206 funding. But the absence of this intervention begs the question: why didn't local and state creek-restoration advocates put more energy into this congressional intervention and a search for funding options? The answer lies partially in political and environmental policy schisms that have developed between the Salt Lake City mayor (Rocky Anderson) and the state legislature.

Anderson, now halfway through his second term, has taken a hard line on curbing regional sprawl. In particular, Anderson's veto of a planned regional mall in the city (which he dubbed the "Sprawl Mall") and opposition to a new proposed state highway (sited on wetlands) has made him few friends in the capitol. In response to Anderson's positions on the mall and highway projects, the state legislature has retaliated by withholding funds for city-sponsored initiatives. This rift has therefore made it difficult to garner support from the state legislature for an environmental undertaking such as the City Creek daylighting project. To get around this impasse, Anderson has included an Open Space Bond in his recommended 2007 budget that includes funding for the daylighting effort.

At this point, active community and political support for daylighting City Creek is growing but is still not as widespread as some planners had hoped. There are indications, however, that a local push to get the work completed is now underway.

First, Mayor, Anderson, in a letter to Utah Senator Bennett, asked for the Senator's assistance to arrange the necessary federal funding for the completion of the Feasibility Study.[114] The lack of response may be an indication that federal funds are not forthcoming.

Second, nongovernmental agencies (such as Ducks Unlimited and the Trust for Public Lands) are stepping up their efforts to heighten awareness of an interest in the Daylighting Project.

Third, city planning staff members have arranged tours for the proposed daylighted area for environmental groups, elected officials at various levels, members of national organizations such as the League of Cities and Town.

Fourth, city planners are looking at using URRI funding to develop a virtual tour of the area. Such a virtual tour would be in the form of a computer-enhanced video that would show the Daylighting Project as envisioned. This type of video would replace a three-dimensional table model and could be used in various briefings to garner additional support for the Daylighting Project.

The increasing need for flood-control measures surrounding the entombed creek has Salt Lake City's water utility looking at the proposed daylighting as a solution for excess flows in the two-mile culvert that currently drains the creek into the Jordan River. This could potentially mean an influx of government funds to the project.

A recent agreement signed by the city and Union Pacific Railroad also has given the project new attention and encouragement. This agreement ensures the acquisition of all land necessary for the daylighting of the stream. It remains to be seen whether these efforts can, collectively, succeed in bringing the City Creek daylighting plan to fruition. Increasingly, however, it has become evident that the funding and leadership for this undertaking may need to come from Salt Lake City and its residents rather than from state officials and the Army Corps. This is a lesson we have learned the hard way.

Notes

1. *See* Salt Lake City Department of Public Utilities, *Salt Lake City Watershed Management Plan* 5 (Nov. 1999).

2. *See The City of Salt Lake: Her Relations as a Center of Trade; Manufacturing Establishments and Business Houses; Historical, Descriptive and Statistical* 18 (L. I. Shaw ed., 1890).

3. *See Salt Lake City Watershed Management Plan, supra* note 1, at 5.

4. *Id.* at 5.

5. Author's personal observation.

6. *See* Salt Lake City Planning Division, *City Creek Master Plan* 2 (1986).

7. *See* Edward Tulledge, *History of Salt Lake City* 72 (1886).

8. *See City Creek Master Plan, supra* note 6, at 2.

9. *See The Politics of Water in Utah: Water of Zion* 28 (Dan McCool ed., 1995).

10. *See* Shaw, *supra* note 2, at 9.

11. *See City Creek Master Plan, supra* note 6, at 2.

12. *See* Shaw, *supra* note 2, at 17.

13. *See* Thora Watson, *The Stream That Built a City: History of City Creek, Memory Grove, and City Creek Canyon Park* 3 (1995).

14. *See* Salt Lake City Department of Water Supply and Waterworks, *Water—For You: Welcome to Salt Lake City's First Water-Treatment Plant* 12 (1955).

15. *Id.* at 4.

16. *See Utah State Water Plan, Jordan River Basin, Public Review Draft* 3–18 (Oct. 1996). The Jordan River at 4,200 feet elevation is in center of the Salt Lake Valley, which drains mountains on both sides rising to 11,000 and 14,000 feet. The Jordan River collects the contamination that travels through the ground water. That contamination is historically from mining at the higher elevations and ranching and agriculture at the lower elevations.

17. *See City Creek Master Plan, supra* note 6, at 2

18. *See* McCool, *supra* note 9, at 29.

19. *See* the commemorative plate on the building, 1892.

20. *See Salt Lake City Watershed Management Plan, supra* note 1, at 6.

21. Photos from the Utah State Historic Preservation Office (on file with the author).

22. *Id.*

23. *See Water—For You, supra* note 14, at 3.

24. Photos from the Utah State Historic Preservation Office (on file with the author) show stagnant pools along the right-of-way of the Creek prior to its entombment in 1909.

25. *See City Creek Master Plan, supra* note 6, at 2

26. *Id.* at 2.

27. See Thomas Alexander, *Grace and Grandeur: A History of Salt Lake City* 62 (2001).

28. *See* Environmental Protection Agency, Clean Water Act (1977), *available at* http://www.epa.gov/region5/water/cwa.htm (last visited Sept. 1, 2005): "Growing public awareness and concern for controlling water pollution led to enactment of the Federal Water Pollution Control Act Amendments of 1972. As amended in 1977, this law became commonly known as the Clean Water Act. The Act established the basic structure for regulating discharges of pollutants into the waters of the United States. It gave EPA the authority to implement pollution control programs such as setting wastewater standards for industry. The Clean Water Act also continued requirements to set water quality standards for all contaminants in surface waters. The Act made it unlawful for any person to discharge any pollutant from a point source into navigable waters, unless a permit was obtained under its provisions. It also funded the construction of sewage treatment plants under the construction grants program and recognized the need for planning to address the critical problems posed by nonpoint source pollution."

29. *See* Tim Palmer, *Endangered Rivers and the Conservation Movement* 93 (1986).

30. *Id.* at 95.

31. Author's personal observation

32. *See* Palmer, *supra* note 29, at 95

33. *Id.* at 95.

34. *See* Robert Gottlieb, *A Life of Its Own: The Politics and Power of Water* 276 (1988).

35. *See* Palmer, *supra* note 29, at 233–34.

36. *See* Salt Lake City, *City Creek Section 206 Aquatic Ecosystem Restoration Detailed Project Report* ¶ 1.2 (Dec. 2003) (on file with author).

37. *See* Environmental Protection Agency, *Brownfields Cleanup and Redevelopment* (1995), *available at* http://www.epa.gov/swerosps/bf/basic_info.htm (last visited Sept. 1, 2005).

38. Pub. L. No. 107-118 (2001).

39. *See* Redevelopment Agency of Salt Lake City, *What Is a Brownfield?*, available at http://www.slcgov.com/CED/RDA/First%20Level/Brownfields.htm (last visited Sept. 1, 2005).

40. Interview with Doug Dansie AICP, Salt Lake City Planning Division (Feb. 2002).

41. *Id.*

42. *See* Salt Lake City, *Gateway and Brownfields Resource Center*, available at http://www.ci.slc.ut.us/ced/rda/Brownfields (last visited Sept. 1, 2005).

43. U.S. EPA Region 8, under the Inter governmental Personnel Act assignment, provided an EPA scientist to the Salt Lake City Redevelopment Agency for the oversight of the Gateway Brownfields Showcase Project. This agreement also provided up to $900,000 in assistance to the project from 1996 to 2002. *See* Assistance ID No. BP-99860601-3 on file with EPA Region 8, 999 18th Street, Suite 500, Denver, CO 80202-2466.

44. *See* Funding Request from Stephanie Wallace to Tom Rogan, Chairperson, RDA Board of Directors Salt Lake City, Utah (Mar. 3, 2000).

45. *See Creating an Urban Neighborhood, Gateway District Land Use and Development Master Plan* 3 (undated).

46. Interview with Valda E. Tarbet, Deputy Director, Salt Lake City Redevelopment Agency (Sept. 2004).

47. *Id.*

48. *See* Center for Brownfields Initiatives, *EPA Region 8*, available at http://www.brownfields.com/Feature/Feature-Awards2003-region8.htm (last visited Sept. 1, 2005).

49. *See City Creek Section 206, supra* note 36, at ¶ 3.5.2.

50. *Id.*

51. See Salt Lake City Zoning Map.

52. *See City Creek Section 206, supra* note 36, at ¶ 3.5.2.

53. *Id.*

54. *Id.*

55. Clare Brandt, *Communities Access the Bonneville Shoreline Trail*, The Trust for Public Lands Utah, Fall/Winter 2004, at 5.

56. Author's observation.

57. South Pacific Division Sacramento District, USAGE, *Section 206 Fact Sheet* (May 2001).

58. *See City Creek Section 206, supra* note 36, at ¶ 5.1.1.

59. *Id.* at ¶ 4.2.

60. *Id.* at ¶ 3.1.

61. *See* Funding Request from Stephanie Wallace to Tom Rogan, *supra* note 44.

62. *See City Creek Section 206, supra* note 36, at ¶ 1.2.

63. *Id.*

64. *Id.*

65. E-mail from Scott Miner, Ecosystem Restoration Specialist, USACE, to Scott Stoddard (Mar. 17, 2003) (on file with author).

66. *See* U.S. Fish and Wildlife Service *Comment on City Creek Section 206 Habitat Improvement Plan* (undated) (on file with author).

67. The acknowledgement page of the January 2005 draft of the Euclid SAMP prepared by Mark Leese of URS in Denver lists twenty six members of the Euclid Neighborhood as contributing to the development of the draft.

68. U.S. Army Corps of Engineers, Sacramento District *City Creek Feasibility Study Notes* (Dec. 2003) (on file with author).

69. One indication of the blight in the area is the designation of part of the area as a redevelopment target area. State law requires blighted conditions as a prerequisite for such designation.

70. See SLC Contract No. 08-1-04-0319, § 2.b (Recorded Apr. 7, 2004).

71. Interview with Valda E. Tarbet, *supra* note 46.

72. See *Salt Lake City Corporation v. Union Pacific Railroad Company*, 02:01-CV-655ST (D. Utah 2002).

73. Interview with Valda E. Tarbet, *supra* note 46.

74. Telephone interview with Bradley Hubbard, Army Corp Project Manager (Feb. 2004).

75. *Id.*

76. *Id.*

77. *Id.*

78. *Id.*

79. *See* Letter from Ross C. Anderson, Mayor of Salt Lake City, Utah, to Hon. Robert Bennett, Utah Senator (Jan. 20, 2005) (on file with author).

80. *See* www.usace.army.mil (last visited Sept. 1, 2005).

81. E-mail from Scott Miner, *supra* note 65.

82. Army Corps of Engineers, *Water Resources Policies and Authorities Ecosystem Restoration: Supporting Policy Information*, CECW Regulation No. 1165-2-502 (Sept. 30, 1999).

83. *Id.* 16.

84. CECW Regulation No. 1165-2-502, *supra* note 82.

85. Interview with Valda E. Tarbet, *supra* note 46.

86. *See* Scott Stoddard & Ron Love, *Buried beneath Downtown: Daylighting Salt Lake City's City Creek* (slide presentation) (2004) (on file with author).

87. Interview with Valda E. Tarbet, *supra* note 46.

88. *Id.*

89. The RDA set aside $900,000 for the development of the project. The RDA also provided funding toward a Rail Relocation Study and funding to augment the funds available from the Army Corps for the geotechnical and environmental studies of that portion of the project that falls within the redevelopment target area. *See* U.S. EPA Cooperative Agreement (Aug. 22, 2000).

90. *See Draft Interim Environmental Sampling and Analysis Report for the City Creek Daylighting Project, Salt lake City, Utah* (MSE Millennium Science and Engineering, Inc., Mar. 3, 2003).

91. *Id.*

92. *Id.*

93. Risk-based concentrations define acceptable levels of concentrations of certain analytes based on exposure limits.

94. *Id.* at 9.

95. *Id.* at 12.

96. *See* Environmental Protection Agency and Department of the Army, *Restoration of Degraded Urban Rivers*, Memorandum of Understanding (July 2, 2002), *available at* http://www.epa.gov/swerrims/landrevitalization/download/epa-usace_urban_water_mou.pdf (last visited Sept. 1, 2005).

97. *Id.*

98. Jonathan Deason, *Urban River Restoration Initiative: Key to Brownfields Redevelopment Success in Urban River Corridors* 2 (Sept. 2001), *available at* http://www.gwu.edu/~eemnews/spring2001-images/PDFVersionofBF2000Paper.pdf (last visited Sept. 1, 2005). The eight URRI pilot projects include (1) Anacostia River in the District of Columbia and Maryland (a Brownfields, freshwater restoration project), (2) Blackstone-Woonasquatucket Rivers in Rhode Island and Massachusetts (a project to clean up contaminates "threatening human health, wildlife, fish habitats and recreational fishing"), (3) Elizabeth River in Virginia ("contaminated by heavy metals from military and industrial sources that pose threats to human health and wildlife"), (4) Tres Rios in Arizona (a riparian habitat restoration project), (5) Passaic River in New Jersey (a seven-mile stretch of river encompassing four counties; project stresses intergovernmental and private agency coordination), (6) Gowanus Canal and Bay in New York ("includes approximately 130 acres of open water" that is impacted by sewer outfalls), (7) Fourche Creek in Arkansas ("located within an EPA Brownfields Assessment Demonstration Pilot"), and (8) City Creek/Gateway District, Utah ("encased in 1910," the creek is to be daylighted through "a 1.5-mile stretch of railroad that traverses a residential, commercial and light industrial area of the city—an area in transition").

99. According to an EPA Region 10 fact sheet on the subject, total maximum daily load (TMLD) is a tool for implementing water-quality standards and is based on a relationship between pollution sources and in-stream water-quality conditions.

100. *See* Deason, *supra* note 98, at 2.

101. *See Restoration of Degraded Urban Rivers, supra* note 96.

102. Remarks by Jane Merger, USACE HQ2, at Urban Rivers Forum Meeting, June 22 and 23, 2004, Salt Lake City, Utah (June 2004).

103. *Id.*

104. The URRI coordinator for EPA Denver reported the addition of $50,000, thereby doubling the money committed to the Euclid SAMP. Telephone interview with Judith McCulley, URRI Coordinator, EPA Denver (Nov. 2004).

105. *See Restoration of Degraded Urban Rivers, supra* note 96.

106. *Id.*

107. E-mail from Beverley Getzen Chief, Office of Environmental Policy, USACE (Dec. 2004) (on file with author).

108. Salt Lake City uses master plans for planning large areas of the city and SAMPs to do detailed planning focused on smaller areas.

109. URRI grant application submitted by Salt Lake City, Utah, to EPA Denver (Aug. 2003).

110. *See* Superfund Technical Assessment and Response Team, -2 US EPA Contract No. 68-W-00-118, Euclid Small Area Master Plan URS Operating Services, Inc. (Dec. 2004).

111. The Salt Lake City Planning Division held an initial meeting to announce the undertaking of a comprehensive review of the Euclid Neighborhood on May 27, 2004, at the City Front Apartment's Lobby, 631 West North Temple Street, Salt Lake City, Utah.

112. The acknowledgment page of the January 5 draft of the SAMP prepared by Mark Leese of URS in Denver lists twenty-six members of the Euclid neighborhood as contributing to the development of the draft.

113. The author has been involved in the development of a significant number of master plans for various areas of Salt Lake City over the past seventeen years, including the master plan for the eastern half of Salt Lake City (which the author was involved with from 1997 to 2004). One of the biggest challenges faced was getting consistent community involvement. A maximum of ten community residents remained involved in the process through the City Planning Commission presentation. In the author's experience, this is normal. The sustained involvement of a significant number of individual property and business owners and managers for the Euclid SAMP is highly unusual.

114. *See* Letter from Ross C. Anderson, *supra* note 79.

6

Bankside San Jose

Richard Roos-Collins

The Guadalupe River originates in the Santa Cruz Mountains and flows northwest through San Jose, California, into San Francisco Bay.[1] Since statehood in 1850, the river has been extensively developed for water supply, flood protection, residential and commercial facilities in the floodplain, and other economic uses.[2] Even though it is located in the heart of Silicon Valley, it remains a spawning and rearing habitat for Central Coast steelhead and Chinook salmon, which are coldwater anadromous species, and for warmwater fish as well.[3] Its banks are riparian habitat for many wildlife species, including foxes, possums, ospreys, and frogs.[4] The river is popular for many forms of recreation, such as seasonal boating, hiking, and picnicking at the several public parks that permit access along the banks.[5] This urban stream is now the locus of a collaborative experiment in restoration managed to enhance economic uses.

The Santa Clara Valley Water District, the local agency responsible for water supply and flood protection,[6] is committed to measures worth more than $250 million to restore to good condition natural resources of the Guadalupe (and two adjacent streams) degraded by nearly 150 years of urban development.[7] The Water District will study, construct, and manage these measures in cooperation with the Guadalupe-Coyote Resource Conservation District (GCRCD), federal and state regulatory agencies, and other parties.[8] This restoration program, which results largely from settlements described in this chapter, will include enforceable objectives, rigorous monitoring of environmental conditions, and adaptive management of the individual measures to ensure accountability for the promised results.[9]

This chapter examines how the Water District, GCRCD, and other parties developed a joint scientific record as the basis for their negotiations and how the resulting settlements use adaptive management to

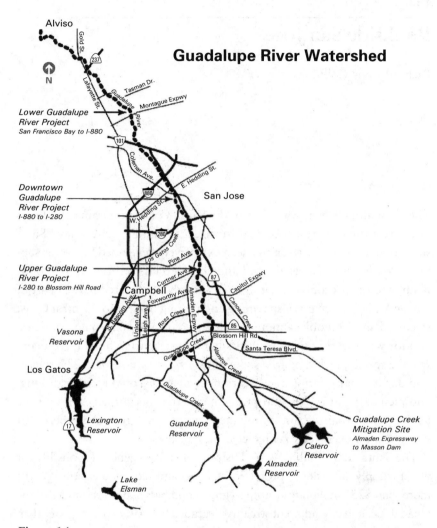

Figure 6.1
Map of the Guadalupe River watershed, 2005. Reprinted with permission of the Santa Clara Valley Water District.

ensure cost-effective restoration in the face of continuing uncertainty about the impacts of water supply and flood-protection facilities. First, it addresses the settlement of a water-rights complaint brought against the Water District to modify the operation of its water-supply system in the upper reach of the river. It then explores the settlement of a related notice of citizens' suit brought against flood-protection projects in the more urbanized reaches downstream. The chapter concludes with a discussion of the future implementation of these settlements, including consequences for both the Guadalupe watershed and other urban rivers.

Water-Rights Settlement

The Guadalupe-Coyote Resource Conservation District is a special local district that advises landowners in central San Jose on best management practices for their lands and other natural resources.[10] In July 1996, GCRCD, joined by Trout Unlimited and by the Pacific Coast Federation of Fishermen's Associations as nonprofit allies,[11] filed an administrative complaint alleging that the Santa Clara Valley Water District holds and uses water rights to store and divert flows in a manner that causes unlawful harm to the coldwater fisheries and other natural resources of the Guadalupe River and two adjacent streams, Coyote and Stevens.[12] GCRCD brought the complaint before the State Water Resources Control Board, which has exclusive jurisdiction to issue or amend appropriative water rights initiated subsequent to 1914.[13] The complaint sought to apply to an urban stream the precedent of the Mono Lake cases, which conditioned the Los Angeles Department of Water and Power's rights to divert tributary inflow to protect the public trust in Mono Lake, located in the remote and rural Eastern Sierra mountains.[14] However, this complaint was resolved by negotiation.[15] The resulting settlement is an important precedent because it establishes a joint venture between a water utility and other stakeholders in the perpetual restoration and adaptive management of an urban stream.[16]

The Lawsuit
The Guadalupe-Coyote Resource Conservation District's Complaint
As alleged in the GCRCD's complaint, the Santa Clara Valley Water District holds eight water-right licenses, issued between 1941 and 1985, for storage and diversion of surface flows from the Guadalupe River and

its tributaries for municipal and industrial water supply in Silicon Valley.[17] It operates five dams for that purpose in this watershed.[18] None of these licenses requires a release of minimum instream flow for protection of public-trust resources.[19] Certain rivers that reach below these points of diversion run dry in most years from late spring (when the rainy season ends in the San Francisco Bay Area) through late fall (when the rainy season begins again) because the diversion covers all natural inflow.[20] GCRCD alleged that this use fundamentally alters the historical condition of this river that, as sustained by the aquifer during the dry season, had continuous flows that attracted the original Spanish missionaries in 1797 and subsequent immigrant farmers in the 1800s.[21] It alleged that the use of these rights threatens to extinguish the anadromous fisheries, which depend on continuous flows in the late fall for spawning habit.[22] Such use has degraded habitat for other fish and wildlife species, boating, and other noneconomic uses of the Guadalupe.[23] The complaint alleged that this use of the licenses violates sections of the California Fish and Game Code and the California Water Code, as well as the public-trust doctrine.[24]

The complaint acknowledged that the Guadalupe suffers from the tragedy of the commons[25]—the cumulative impact of 150 years of urban development.[26] Many forms of development, including the permitting of residential and commercial facilities in the immediate floodplain, are wholly outside of the Water District's control.[27] However, the complaint alleged that the Water District is responsible for any degradation caused by its management of its water supply and flood-protection facilities, which largely regulate the river's flows (subject only to minor additional impacts by third parties).[28] GCRCD sought to hold the Water District accountable only for the proportional impacts of its own facilities.[29] In effect, the complaint relied on an 1884 case, which was the first in California to apply the public-trust doctrine to impairment of navigable waters.[30] In *Gold Run Ditch*, the California Supreme Court prohibited hydraulic mining that, as undertaken by a multitude of individual miners, had resulted in discharges of soil and other debris into nonnavigable tributaries, eventually impairing navigation in the Sacramento River:[31]

As a navigable river, the Sacramento is a great public highway, in which the people of the State have paramount and controlling rights. These rights consist chiefly of a right of property in the soil, and a right to the use of the water flowing over it, for the purposes of transportation and commercial intercourse. . . . To make use of the banks of a river for dumping places, . . . is an encroachment upon the soil of the latter, and an unauthorized invasion of the rights of the public to its navigation; and when such acts not only impair the navigation of a river, but at the same time affect the rights of an entire community or neighborhood, or any considerable number of persons, to the free use and enjoyment of their property, they constitute, however long continued, a public nuisance.[32]

While the miners had acted independently and separately and while their individual actions may have been "slight" or "scarcely appreciable," the "common result" was impairment of navigation on the Sacramento River. Accordingly, they were jointly and severally liable for the public nuisance and subject to a "coordinate remedy."[33]

The complaint requested that the State Water Board adopt several remedies, following public notice and hearing.[34] These were (1) a disclosure of the operating protocols of the Water District's water-supply facilities, including the quantities and schedules of its diversions relative to natural inflows; (2) a cooperative investigation of the impacts of these facilities on the coldwater fisheries and of alternatives to mitigate any adverse impacts; (3) following such investigation, amendments to the water-rights licenses to include flow schedules adequate to maintain the coldwater fisheries and other public-trust resources in good condition; and (4) further amendments to require a program of nonflow measures to restore the channel form and riparian vegetation of the river.[35] Such measures complement a flow schedule to restore the quantity and quality of fish habitat and may include placement of spawning gravel, planting of trees, and removal of structures that block fish passage either upstream or downstream.[36]

The Santa Clara Valley Water District's Answer   The Santa Clara Valley Water District answered the Guadalupe-Coyote Resource Conservation District's complaint in October 1996.[37] The answer stated generally that the status quo "presently presents the appropriate balance of competing needs and interests"[38] and requested dismissal of the complaint.

The Water District's answer described the purpose and benefits of the water-rights licenses.[39] The Water District, including its predecessors, has been responsible since 1929 for conserving surface waters and groundwaters and for importing additional waters, as appropriate for the supply of Santa Clara County, which encompasses 1,300 square miles.[40] It serves thirteen local districts and companies, which deliver water to the taps of 1.6 million residents from San Jose northward up the San Francisco Peninsula.[41] Appropriations from local streams, as well the import of an even greater amount of water from the State Water Project and the federal Central Valley Project,[42] are necessary to ensure adequate water supply and to prevent land subsidence. Such subsidence had occurred in the 1800s through the early 1900s as a result of continuous groundwater overdraft.[43] The land surface sank up to fifteen vertical feet in some locations as the hydrostatic pressure of the aquifer (namely, the vertical force of such water to hold up the soil) was depleted.[44] Such subsidence had threatened the safety of residential and commercial facilities, saltwater intrusion from the San Francisco Bay, and storage capacity of the aquifer.[45] Today, the aquifer is stable as a result of the Water District's program of regulated pumping and also deliberate percolation of surface flow via spreading ponds back into the aquifer.[46] The District's answer also emphasized that another second statutory function, flood protection, allows conservation of peak flows from the Guadalupe and other local streams for water supply.[47] In sum, the Water District "has implemented a comprehensive water operations strategy that has resulted in a fully integrated water supply system."[48]

The Water District further stated that reservoir parks on the Guadalupe and other local streams are popular for recreation and provide substantial habitat for warmwater fish and wildlife.[49] It alleged that releases of minimum flows may cause significant harm to the water supply as well as noneconomic uses of the reservoirs[50] and that the benefits of such releases for the downstream coldwater fisheries and other resources are unknown or at least unproven in the GCRCD's complaint.[51]

The Water District objected legally to GCRCD's claims. It alleged that Fish and Game Code section 5937 applies in mandatory form only to licenses in the Eastern Sierra[52] or, in the alternative, only to permit or license applications filed after 1975 when the State Water Board adopted a rule applying section 5937 prospectively throughout the state.[53] The Water District stated that it actively cooperates with the California Department of Fish and Game, which has primary authority to enforce this and other sections of the Fish and Game Code, and that Fish and Game has not requested any minimum flow schedule, fishway, or other measure not in place.[54]

The Water District argued that, in the absence of a mandatory duty to amend the licenses, the State Water Board may undertake a discretionary balancing of the public interest under relevant state laws.[55] It argued that any such balancing must take into account various factors that favor the status quo, including the economic viability of Silicon Valley, the potential waste of water in the absence of scientific evidence determining what minimum flow release at a given facility would restore the downstream coldwater fisheries to good condition, potential adverse impacts by such releases to reservoir uses, contributions of many third parties to the existing conditions of the fisheries, including barriers to fish passage and flow diversions, and the reliance on the licenses, which, as issued, do not require such minimum flow releases.[56] Finally, the Water District argued that GCRCD had unclean hands, having tolerated these operations over the course of many decades.[57]

The State Water Board did not set the GCRCD's complaint for hearing or permit further briefing. Instead, in October 1997, the Water District and the Fish and Game Department proposed to undertake the Fisheries and Aquatic Habitat Collaborative Effort (FAHCE) to resolve the complaint.[58] Other regulatory agencies with jurisdiction over these streams and GCRCD agreed.[59] While the motives varied and are confidential, the Water District and other stakeholders[60] faced substantial expenses and uncertain odds in litigation, given the novelty of many of the claims.[61] Each stakeholder also recognized the potential that a settlement would create mutual gains not otherwise achievable—for example, by including measures that the State Water Board would not order in a disputed hearing of the complaint.[62] An example is an adaptive management program, which commits the Water District and other stakeholders to joint implementation of restoration measures.[63] The State Water Board cannot order a nonlicensee to make such a commitment because, under the Water Code, it does not have personal jurisdiction over any entity that does not hold a water right, but it may accept the commitment as made in a settlement with a licensee.[64]

The Fisheries and Aquatic Habitat Collaborative Effort Process

The Water District and Fish and Game Department proposed a specific structure for collaborative process.[65] The parties refined and adopted this process in organizational meetings through early 1998 and then implemented it through January 2003 when they entered into settlement.[66] The process had six features that proved to be critical to its eventual success.[67]

First, the negotiating table was larger than the Water District, Fish and Game Department, and GCRCD.[68] It included other agencies whose support will materially affect whether the State Water Board approves the settlement as the basis for amending the water-right licenses.[69] The U.S. Department of Commerce's National Marine Fisheries Service, which manages, conserves, and protects living marine resources that spend at least part of their life cycle within the U.S. exclusive economic zone,[70] will be responsible for ensuring that the settlement complies with the Endangered Species Act,[71] which protects the threatened steelhead fishery[72] against take.[73] The U.S. Fish and Wildlife Service,[74] which conserves, protects, and enhances fish, wildlife, and plant resources that do not use marine habitat or otherwise are not under National Marine Fisheries Service's jurisdiction,[75] will ensure that the settlement complies with the Fish and Wildlife Coordination Act.[76] The San Francisco Regional Water Quality Control Board will advise the State Water Board whether the settlement complies with the water-quality standards adopted by the Basin Plan for the Guadalupe.[77] It participated in the negotiations as an advisor to the other stakeholders.[78] Its formal participation might have constituted a predecisional commitment, since it is a subdivision of the SWRCB that will decide whether to approve the settlement as license amendments.[79] Finally, the City of San Jose participated for several reasons. It operates several water-control facilities under its own licenses; a stormwater drains and collection system, which discharges some stormwater back into the streams; and the wastewater treatment facility, which is a potential source of recycled water[80] for reuse in a minimum-flow schedule. It administers land-use laws applicable to the floodplain of the Guadalupe and the other streams included in the negotiation.[81] It also has a general duty to protect the public welfare of the residents, including development of improved recreational access and facilities.[82]

Second, the collaborative process had a single purpose: development of a management plan that, as applied to the Water District's facilities and operations on the Guadalupe and other streams, will ensure compliance with all laws that require protection of the coldwater fisheries and other trust resources.[83] The plan will include "innovative solutions for improving fisheries habitat in the County which provide cumulative benefits for the community.[84] For example, the plan will consider collaboration with the City of San Jose's proposal for streamflow augmentation with recycled water as part of this effort."[85]

Third, the stakeholders jointly interviewed and selected a neutral facilitator to schedule and manage all subsequent meetings.[86] Although the Water District paid the facilitator's fees and related meeting expenses, the consulting contract clearly provided that the duty of loyalty ran to the process only and that the resulting process management would be consensual.[87]

Fourth, the stakeholders established two standing committees to undertake the hard work of developing the management plan.[88] The Technical Advisory Committee consisted of technical staff responsible for collection, review, and analysis of all scientific data relevant to understand the Water District's impacts on public trust resources, both under today's baseline conditions and under alternatives that may mitigate existing impacts.[89] A Consensus Committee consisted of managers responsible for negotiating the management plan and taking into account the economic, social, and legal merits of the alternatives that the Technical Committee found to be technically feasible to mitigate adverse impacts on the fisheries.[90]

Fifth, an expert fisheries consultant assisted the Technical Committee to develop and implement a Limiting Factors Study.[91] As with the facilitator, the parties jointly selected the consultant. While the Water District then entered into a consulting contract to pay his fees and expenses, his duty of loyalty ran solely to the Consensus Committee.[92] The Limiting Factors Study was intended to identify and rank all physical conditions (such as water temperature, presence of spawning gravels, barrier to fish passage, or presence of riparian cover) that affect the population or distribution of the coldwater fisheries in the streams; for each limiting factor, identify the proportionate contribution of the Water District's facilities relative to third parties'; and identify and evaluate for technical feasibility the flow and nonflow measures that might improve fisheries habitat by mitigating the Water District's existing impacts.[93]

Sixth, the stakeholders agreed to start with the Limiting Factors Study, then negotiate on the basis of that scientific record, and conclude the process in three years.[94] The State Water Board stayed its hearing on the

GCRCD's complaint.[95] The parties subsequently extended that deadline to January 2003 to permit additional study of the coldwater fisheries, whose life cycle is more than three years.[96] Still, the deadline motivated the stakeholders to make a disciplined effort to resolve issues expeditiously.[97] Any extension required mutual consent and assurance of continued commitment to keep the shoulder to the wheel.

The Technical Committee and consultant undertook three years of field studies, including surveys of the physical form of streambed and banks, electrofishing to establish population counts by reach, and flow and temperature monitoring on a continuous basis.[98] In March 2000, it completed a Limiting Factors Study.[99] The study summarized existing scientific literature relevant to the stated purpose, mapped the existing habitat conditions of each stream reach affected by the Water District's facilities, analyzed the impact of each of eleven limiting factors by reach, parsed the contributions of the Water District and third parties to such impact, and recommended alternatives for mitigation of adverse impacts.[100]

Many study findings were inconsistent with parties' expectations based on personal observations before the study. For example, the study reported the known fact that a reservoir in this watershed, warmed by the Mediterranean climate, develops a thermal stratification each summer, whereby surface water exceeds 70 degrees Fahrenheit while deeper water is much cooler.[101] The study found that that stratification has a significant and previously unknown consequence for the resolution of the complaint: the rate of minimum-flow release will determine the continued availability of coldwater in a given reservoir as the summer progressed.[102] A higher release schedule will deplete such availability quicker and thus will subject downstream fish to more but warmer flows potentially unsuitable for their spawning. While the study does not purport to be definitive, the Technical Committee jointly recommended its use, and the Consensus Committee used the study findings to guide negotiation.[103] Thus, the Consensus Committee used a joint scientific record as the basis for choosing among measures to include in the eventual settlement.[104]

Negotiation effectively began on receipt of the Limiting Factors Study. Since negotiations of litigation are confidential,[105] this chapter reports only the protocol used to develop, draft, and refine concepts into the form of settlement.

The Consensus Committee used a protocol known as "one-text draft-ing."[106] This mitigates against the risk or fear that the defendant in a water or other environmental resources case will unduly control a collaborative process because it has disproportionate resources. Under this protocol, any party may volunteer to prepare a first draft of a given document.[107] Other parties will comment in advance of the next meeting. The preferred form of comment is "yes," "no," or "yes if. . . ." Parties will discuss comments and seek to resolve disputes at the next meeting.[108] A party other than the initial drafter will then prepare the second draft, showing proposed changes reflective of meeting discussion in redline/strikeout format.[109] The process will continue in this seriatim manner.[110] At any given meeting, only the latest draft is on the table for review.[111] The Consensus Committee effectively used this protocol to draft and negotiate more than a dozen drafts until all parties approved the final settlement in January 2003.[112]

Fisheries and Aquatic Habitat Collaborative Effort Settlement

The FAHCE settlement states its purpose as resolving all claims in the Guadalupe-Coyote Resource Conservation District complaint and all issues relating to the Santa Clara Valley Water District's compliance with other federal and state laws applicable to its water-supply facilities, ex-cepting only a natural resources damages claim relating to the capture of mercury leachate that originated in New Almaden Mine, located upstream of the water-supply facilities.[113] The settlement consists of con-tractual provisions stated in articles I to V and IX to X (which establish how the settlement will be used in the State Water Board's proceeding to amend the licenses and related regulatory proceedings) and flow and non-flow restoration measures stated in articles VI to VIII (which are proposed for incorporation into the licenses for subsequent implementation).[114]

The contractual provisions manage the necessary but awkward reality that the parties that are public agencies entered into a settlement in ad-vance of the preparation of an environmental document required by the California Environmental Quality Act[115] for the State Water Board's approval and any other state action on the settlement and by the National Environmental Policy Act[116] for any federal action, such as the Biological Opinion required by the Endangered Species Act, section 7.[117] The settlement balances the support for the agreed-to restoration measures against the agencies' duties not to bind themselves in advance of such an environmental document and consideration of public comments.[118] The settlement represents that the parties concur, on the basis of the Limiting Factors Study and other evidence in the existing record, that these restoration measures will comply with all applicable laws.[119] It provides that these measures will be the project[120] for review in the environmental document. It further provides that the parties will support all necessary approvals of these measures without substantial modification,[121] unless the public record as subsequently developed demonstrates that another alternative will better protect and maintain the beneficial uses of these waters.[122] In that event, the parties will consider potential amendments to the settlement pursuant to a dispute resolution procedure.[123] Assuming that the settlement is approved without substantial modification, GCRCD will dismiss its complaint.[124] Following such approval, the Water District will implement the measures as incorporated as license amendments.[125] The parties will not seek to reopen the licenses or the underlying settlement, unless significant new information (including change in applicable law) materially changes the bargained-for benefits.[126] The term of the settlement is perpetual,[127] unless terminated due to the Water District's withdrawal following compliance with the dispute-resolution procedure.[128] The settlement is a contract enforceable by specific performance as a supplement to any remedy for enforcement of the licenses under general laws.[129] As of the settlement's publication date, the parties anticipate that the State Water Board will take final action on the settlement by mid-2008.[130]

The settlement establishes a perpetual program for restoration of the local streams that the Water District uses for its water supply, including the Guadalupe.[131] This program has several fundamental parts.

The settlement provides that the overall management objectives are to restore and maintain steelhead and salmon fisheries in good condition in each stream.[132] It provides that an Adaptive Management Team, which includes all signatories,[133] will restate these qualitative objectives in a measurable form for the purpose of monitoring and adaptive management.[134] Examples of such objectives are an amount of spawning gravel in a given reach or the percentage of the water surface that should be shaded by riparian vegetation to maintain coldwater. These objectives will be enforceable conditions of the water right licenses.

The Water District will release a minimum flow from each reservoir or diversion facility.[135] The release schedules, which vary across the reservoirs and watersheds, are intended to maximize the geographic extent and duration of coldwater flow for spawning and rearing.[136] In the Guadalupe River watershed, the release schedules are stated not in traditional form (as a value in cubic feet per second) but instead as an obligation to implement a rule curve for each reservoir to maximize the coldwater habitat, taking into account a given year's hydrologic, weather, and other circumstances.[137] The Water District will follow a ramping rate to temper any artificial change in flow release.[138] In addition, it will undertake further study of the feasibility of delivering recycled water from the City of San Jose's wastewater treatment facility near San Francisco Bay back uphill to the local creeks or of managing the stormwater collection system for the same purpose and will implement such measures found to be feasible and suitable.[139]

In addition to the flow measures, the Water District will construct, operate, and maintain nonflow measures in four phases.[140] In phase one, which will begin on the effective date and continues for ten years,[141] it will remove certain weirs (namely, bank-to-bank structures used to raise the vertical height of flow without substantial storage) and other low barriers to fish passage.[142] The Limiting Factors Study identified each such barrier and assigned a priority based on the feasibility of removal and the significance of the currently unavailable habitat.[143] In phase two (years eleven to twenty), the Water District will remove other barriers, either directly or by contribution if owned by third parties.[144] It may also implement a trap-and-haul program to transport spawning adults to habitat blocked by storage dams, if necessary to achieve the management objectives.[145] Phase three (years twenty-one to thirty) continues that same obligation.[146] In phase four (years thirty-one to perpetuity), the Water District will continue to maintain all nonflow measures constructed in prior phases.[147]

Finally, in consultation with the Adaptive Management Team, the Water District will implement these obligations in an adaptive manner.[148] In phase one, it will develop a fish habitat-restoration plan,[149] including a geomorphic functions study,[150] to specify the locations and other details of nonflow measures. It will develop operation and maintenance procedures, more detailed forms of the rule curves in Settlement appendix E, for the flow measures.[151] The plan will include measurable objectives to implement the qualitative management objectives. In continuing collaboration with the Adaptive Management Team, the Water District will systematically monitor the changing conditions of the fisheries as these measures are implemented.[152] It may modify flow and nonflow measures alike if, on the basis of monitoring results, the Adaptive Management Team determines that modifications will better contribute to timely achievement of the management objectives.[153] It will spend up to $42 million in each of phases one, two, and three and whatever amount is necessary thereafter to continue the flow and nonflow measures already implemented.[154]

The Integrated Guadalupe Flood-Control Settlement

The Guadalupe is a small urban river. Its average flow is forty eight cubic feet per second (c.f.s.).[155] During the rainy season from November through April, its peak flow may be several orders of magnitude more.[156] The 100-year flood (a flow predicted to occur once a century) is 17,000 c.f.s.[157] Large floods have occurred many times since statehood in 1850.[158] Today, more than 3,000 homes and 1,000 commercial and industrial buildings, including many of the premier computer companies of Silicon Valley, are located in the 100-year floodplain, which includes the riparian and valley lands above the river channel into which such flood flows would spill absent intervention.[159]

The Santa Clara Valley Water District is the local agency that provides flood protection,[160] while Santa Clara County and municipalities permit land-use developments.[161] As in most urban watersheds in California or the nation, it has always been and is legal under local ordinance to permit developments in the floodplain.[162] As a result, the Water District must intervene systematically to redirect flood flows as necessary to protect life and property. Its plan of flood protection in this watershed consists of three projects. The Upper Guadalupe Flood Control Project (FCP) begins in the foothills of the Santa Cruz Mountains and continues downstream or northward to Interstate 280. The Downtown Guadalupe Project begins at Interstate 280 and ends at Interstate 880. The Lower Guadalupe Project begins at Interstate 880 and continues to the town of Alviso, near San Francisco Bay.

The planning, financing, and construction processes for these projects are complex regional efforts that have spanned five decades and counting. The lower and downtown projects became operational in January 2005, and the upper project is still under preliminary construction.[163] The design and operation of these projects will be integrated as a result of recent settlements to contribute to the restoration of the coldwater fisheries in the Guadalupe and recreational enhancements, including trails, parks, and other forms of public access.

Notice of a Citizens' Suit against the Downtown Flood Control Project

In 1986, Congress authorized the United States Army Corps of Engineers, in partnership with the Santa Clara Valley Water District and the City of San Jose, to construct the Downtown Guadalupe Project.[164] In February 1992, the Regional Water Quality Board issued the final regulatory approval, which set forth water-quality certification[165] and waste-discharge requirements.[166] The approved project consisted of hardscape (such as concrete armoring and training walls) in the river's channel as necessary to increase the hydraulic capacity from the existing 8,000 c.f.s. to 17,000 c.f.s.[167] This certification required mitigation measures to protect aquatic habitat, including development of a Mitigation and Monitoring Plan, planting of riparian vegetation, and maintenance of a low-flow channel for fish passage outside of the flood season.[168] The certification also included an obligation to assist in the implementation of the City of San Jose's River Master Plan for recreational facilities and access.[169] That plan, as developed in the 1980s, provides for a linked complex of gardens (including several dedicated to heritage roses and Sister Cities), a visitor's center, tennis courts, and riparian trails.[170]

The Army Corps and the Water District completed the lower reaches (called Contracts 1 and 2) by 1996.[171] These reaches, located in the flight path of San Jose International Airport, were largely undeveloped. The upstream Contract 3 is more urbanized: its banks are already occupied by a complex maze of freeway and railway bridges, buildings, and other developments.[172] In May 1996, before construction of Contract 3 began, the Guadalupe-Coyote Resource Conservation District issued a notice of citizens' suit under Clean Water Act section 505[173] to enforce the 1992 certification.[174] The notice named the Army Corps and the Water District as project sponsors.[175] It alleged that the Mitigation and Monitoring Plan required by the 1992 certification had not been fully approved by resource agencies and that such approval was a condition precedent for construction of Contract 3.[176] It alleged that some mitigation measures constructed in Contracts 1 and 2 did not comply with the performance requirements of the 1992 certification or underlying water-quality standards and had already failed in minor floods.[177] The notice proposed negotiation, while stating that GCRCD would seek damages, injunctive relief, and attorneys' fees in any litigation in U.S. District Court.[178]

The Water District and the Army Corps did not immediately grasp this olive branch. The 1992 certification resulted from many years of negotiation between the project sponsors and resource agencies.[179] GCRCD was a latecomer, from their perspective.[180] They were not pleased that GCRCD, a special local district with advisory authority only, appeared to second-guess the measures approved by the resource agencies that have direct authority to regulate design and operation.[181] Further, project sponsors and the GCRCD had developed a mutual distrust as a result of confrontational letters and meetings preceding the Clean Water Act notice.[182] Finally, the GCRCD filed its water-rights complaint shortly after this notice.[183] The Water District initially viewed this double-whammy as a threat to its flood-protection and water-supply operations in total.[184]

This inertia ended, thanks partly to the initiative of the president of the Guadalupe Parks and Gardens Club. The Parks Club had helped design the riparian parks, which will be features of the downtown project.[185] As a former Assistant U.S. Secretary of Defense, the Parks Club president effectively asked each side: "why is this negotiation so hard to start, if the U.S. can finish nuclear disarmament treaties with the former Soviet Republics?"[186] In June 1997, the resource agencies and project sponsors informally agreed that the mitigation measures required by the 1992 certification should be enhanced in three respects: more on-site planning of riparian vegetation, other measures to prevent warming of water temperature as a result of removal of existing vegetation where necessary to ensure flow capacity, and removal of fish barriers (such as weirs) in the project reaches.[187] GCRCD was invited to join this collaborative process shortly thereafter.[188]

The Water District and the Army Corps did not formally answer the Clean Water Act notice.[189] The notice was eventually withdrawn as a result of settlement, discussed below. The notice is significant as a turning point in integrated management of flood and nonflood flows to enhance the beneficial uses of the Guadalupe.

Downtown Guadalupe Flood Control Project Settlement

The Guadalupe Flood Control Project Collaborative used the process concurrently used in the Fisheries and Aquatic Habitat Collaborative Effort as well.[190] Its purpose was to resolve the Clean Water Act notice in a manner that assured compliance with all applicable laws, including Endangered Species Act section 7, which had become recently applicable as a result of the mid-1997 listing of the Central Coast steelhead.[191] Efforts were divided between a Technical Fact-Finding Subcommittee, which consisted of technical staff, and the Collaborative, which consisted of decisional managers.[192] The project sponsors instructed their environmental consultant, who had been preparing documents related to compliance with the 1992 certification, to undertake further study at the instruction of this Collaborative and to evaluate the hydraulic capacity and cost of various alternative designs for Contract 3 to reduce the project impacts to riparian and aquatic habitat.[193] The Collaborative selected a neutral facilitator, whose fees and expenses were paid by the project sponsors.[194] It set a deadline of July 1, 1998, for settlement. It used one-text drafting as the negotiation protocol.[195]

The Collaborative established criteria to guide the evaluation of alternative designs. It required that, to be approvable, an alternative would provide at least as much flood protection as the current project, achieve measurable objectives for other beneficial uses, result in timely project completion, be cost-effective and fundable, and comply with all applicable laws.[196] Applying these criteria to the studies undertaken in rapid succession by the consultant, the Collaborative unanimously approved a bypass facility that diverted flood flows underground and around a constricted reach of the river channel, as superior to the then-current project that relied on very extensive hardscape of that channel to accomplish the same result.[197] On July 1, 1998, the project sponsors, resource agencies, and GCRCD entered into a settlement in support of that alternative design.[198]

Like the Fisheries and Aquatic Habitat Collaborative Effort settlement, the downtown Flood Control Project settlement was a starting point for

Figure 6.2
Work on the Guadalupe River restoration in downtown San Jose, 2002. Photograph reprinted with permission of the Santa Clara Valley Water District.

regulatory approvals. It proposed a design—two underground culverts each seventeen-feet high and twenty-five-feet wide on the east side of the river in Contract 3—as the preferred alternative for the purpose of environmental review.[199] It required the project sponsors, by April 15, 1999, to develop a Mitigation and Monitoring Plan that provides for replacement of any riparian vegetation that must be removed in certain locations to ensure adequate hydraulic capacity, with new plantings in other locations of equal or superior value for the coldwater fisheries; includes other measures to prevent any harmful increase in water temperature during the transition period when new plantings do not shade the river as well as any removed trees; and provides for adaptive management of the project over its 100-year useful life.[200] The adaptive management consists of measurable objectives for flood protection and environmental benefits, systematic monitoring of actual conditions over time, and (through an Adaptive Management Team consisting of the signatories) modification of project design or operation as appropriate to remedy any deficit.[201]

On April 14, 1999, the parties entered into a supplement to the settlement to confirm that the Mitigation and Monitoring Plan complied with these requirements.[202] The project sponsors then obtained a series of federal and state approvals, concluding with the Regional Water Quality Board's issuance of a new water-quality certification.[203] This certification requires that the Mitigation and Monitoring Plan will be implemented to prevent any net loss in riparian vegetation or other natural values,[204] achieve stated measurable objectives for each beneficial use,[205] and provide for adaptive management of project design and operation by an Adaptive Management Team if, over the project life, the team finds that a measurable objective is not likely to be met.[206]

No stranger to the settlement appealed.[207] Construction of the downtown project concluded in 2005.[208] The GCRCD, while continuing to formally support the project, voices substantial concerns that the as-built design relies excessively on concrete mats, gabions, and other fixtures to maintain channel form and capacity through the downtown area. It believes that the alternative of enhancing the natural channel (to increase hydraulic capacity without the necessity of an underground bypass) was given inadequate consideration during the collaborative process, which plainly was driven by the collective desire to end decades of planning, debate, and litigation. Based on operating experience to date, it also believes that the underground bypass, as well as the concrete mats and other fixtures in the surface channel, will disrupt the capacity of flow to carry suspended sediment and will result in large sediment deposits, thus requiring periodic dredging. It objects that the concrete mats, which include sharp plates and cables, may create substantial hazards to recreation and migrating fish. The Adaptive Management Team will continue to struggle with these and other disputes which go to whether the as-built design is indeed sustainable for flood-control protection as well as other beneficial uses.[209]

Guadalupe Watershed Integration Working Group

In early 2002, at the request of the GCRCD, the Water District established the Guadalupe Watershed Integration Working Group to coordinate the design and operation of the three flood-control projects in this watershed.[210] The lower project began operation in 1985, but the Army Corps and the Water District were revising the design due to inadequate capacity.[211] The upper project was at the end of a planning process preparatory to regulatory approvals and initial construction.[212]

The Watershed Integration Group consists of the same agencies that entered into the settlement for the downtown project.[213] It uses the proven collaborative process.[214] It has an ad hoc Design Review Team, which oversees ongoing technical studies, including collaborative review of environmental documents required for any further regulator approvals.[215] The Watershed Integration Group itself is a policy forum where the negotiators commit to recommend decisions for ratification by their respective directors or boards.[216] Without entering into formal settlements, the Watershed Integration Group developed consensus on the designs, including mitigation conditions, for incorporation into the regulatory approvals for the lower and upper projects.[217] Those approvals have now issued.[218] The lower and downtown projects are operational, and the upper project will be constructed in phases through 2015.[219]

The lower and upper projects include an adaptive management program consistent with the downtown projects.[220] The Water District will undertake specified measures to mitigate impacts on riparian corridor and channel form.[221] The certifications incorporate measurable objectives for environmental results.[222] The Water District will monitor achievement of those objectives and submit annual monitoring reports.[223] The same Adaptive Management Team will evaluate the adequacy of the approved designs to achieve the measurable objective and, within the limits of adaptation approved by the Regional Water Quality Board, adapt the designs (for example, by reconfiguring a levee design) or operations accordingly over the next century.[224] The Water Districts required to undertake further studies in addition to the monitoring programs to refine designs for geomorphic functionality—to ensure that the channel through the affected reaches is capable of handling the water flow and sediment load.[225]

Looking Forward

The Water District will operate its water-supply and flood-control facilities to achieve measurable management objectives for all beneficial uses.[226] It will undertake more than $200 million in physical measures to restore the environmental quality of this stream.[227] It will monitor achievement of the management objectives that state the desired conditions of coldwater fisheries, their habitat, and other natural resources.[228] An Adaptive Management Team, consisting of federal and state regulatory agencies as well as other stakeholders, will collaborate with the Water District to adapt these facilities to achieve these objectives, subject to the constraint that any such adaptation must fall within the scope of the underlying regulatory approvals.[229] The Guadalupe is the locus of a perpetual experiment in maintaining peaceful coexistence of economic and environmental uses of an urban stream.

This effort is a significant precedent for restoration of other urban streams. First, the local district will integrate management of water-supply and flood-protection facilities, even though they were separately permitted and funded, to restore environmental quality. Second, it will be legally accountable for actual results as described by the measurable management objectives. Such accountability is not required by general laws, which provide merely that the permitting agency will predict the foreseeable impacts of a given action. A permit for water use, whether under the Water Code or other substantive law, typically does not incorporate those findings in an enforceable form and thus does not provide for reopener if unexpected impacts occur. Third, stakeholders will participate in a perpetual Adaptive Management Team to cooperate in analysis of monitoring results and any modification in facility design and operation necessary to achieve management objectives. This will require the stakeholders to continue to invest the time and resources necessary to test and potentially modify such design or operation if actual performance differs from expectations.

It is the long-term adaptability and innovative institutional arrangement that make the Guadalupe settlement a model that merits close consideration for other urban watersheds.

Notes

1. *See* Complaint Pursuant to Fish and Game Code Sections 5901, 5935, and 5937; the Common Law Public Trust Doctrine; the Porter-Cologne Water Quality Control Act; and Water Code Section 100 ¶ 40 (1996). The Guadalupe-Coyote Resource Conservation District's complaint concerned California water-right licenses numbers 2205 (Alamitos Creek); 2208 and 2209 (Calero Creek); 2210, 7211, 7212, and 10607 (Coyote Creek); 2206, 2837, and 6943

(Guadalupe Creek); and 5729, 6944, and 11791 (Los Gatos Creek). This chapter focuses only on the Guadalupe River.

The complaint is *available at* www.n-h-i.org/Guadalupe_River.html (1996) (last visited Aug. 5, 2005) or in hard copy from the State Water Resources Control Board, Division of Water Rights, 1001 I Street, Sacramento, CA 95814.

2. Guadalupe Flood Control Project Collaborative, *Record Document* 4–5 (Sept. 1998) ("1998 FCP Settlement Record Document").

3. *Id.*

4. *Id.*

5. *See* Santa Clara Valley Water District (SCVWD), *Fact Sheet: Guadalupe River Park and Flood Protection Project* (2005).

6. *See* Cal. Water Code Appendix §§ 60-1 *et seq.* (2005).

7. The restoration budget is $146 million (2003) for implementation of the water-rights settlement and substantially more than $100 million for the downtown Guadalupe Flood Control Project (*see* SCVWD, *Fact Sheet, supra* note 5).

8. *See* Settlement regarding Water Rights of the Santa Clara Valley Water District on Coyote, Guadalupe, and Stevens Creeks (Jan. 2003) ("FAHCE Settlement"), *available at* www.n-h-i.org/Guadalupe_River.html (last visited Aug. 5, 2005); 1998 FCP Settlement Record Document.

9. *Id.*

10. *See* Cal. Pub. Res. Code §§ 9151 *et seq.* (2005).

11. Trout Unlimited is a conservation group dedicated to the preservation of coldwater fisheries nationwide (*see* www.tu.org). The Pacific Coast Federation of Fishermen's Associations represents commercial salmon fishermen in Western States (*see* www.pcffa.org). California Trout, Inc. joined the settlement (*see* www.caltrout.org). For simplicity, this chapter refers to GCRCD as the plaintiff because it initiated the litigation and had lead responsibility for strategy.

12. Complaint, *supra*, note 1.

13. *See* Cal. Water Code §§ 1250 *et seq.* (2005). The SWRCB regulates other water rights (including pre-1914 appropriative and riparian rights) to prevent waste or unreasonable use. *See id.* §§ 100, 275; Calif. Const. art. X, §. 2.

14. The Mono Lake cases held for the first time that the water rights of an urban water utility in California must be conditioned to protect the public trust in navigable waters, consisting of the uses of fishing, commerce, and navigation. The cases consist of three judicial and two administrative decisions: *National Audubon Society v. Superior Court*, 33 Cal. 3d 419 (1983); *California Trout, Inc. v. State Water Resources Control Board*, 207 Cal. App. 3d 585 (1989) ("*CalTrout I*"); *California Trout, Inc. v. Superior Court*, 218 Cal. App. 3d 187 (1990) ("*CalTrout II*"); and State Water Resources Control Board, Decision 1631 (1994) and Order WR 98-07, *available at* http://www.waterrights.ca.gov (last visited Aug. 5, 2005).

15. FAHCE Settlement, *supra* note 8.

16. *Id.*

17. Complaint, *supra* note 1, ¶¶ 17, 19–27.

18. *Id.* at ¶ 17.

19. *Id.*

20. *Id.* at ¶¶ 54–58.

21. *Id.* at ¶ 44.

22. *Id.* at ¶¶ 54–66.

23. *Id.* at ¶¶ 48–53.

24. The public-trust doctrine requires that a water-rights license must be conditioned to protect public-trust values "whenever feasible." *National Audubon, supra* note 14, 33 Cal. 3d at 446. Such values include commerce, navigation, fisheries, and ecological quality. *Id.* Even where a license makes no provision for release to protect fish and wildlife, a licensee does not have "a vested right to appropriate water in a manner harmful to the interests protected by the public trust." *Id.* at 445. The State Water Resources Control Board has a duty of "continuing supervision" to ensure compliance with this common law. *Id.* at 447. "The case for reconsidering a particular [water-right] decision is even stronger when that decision failed to weigh and consider public trust uses." *Id.*

The public-trust doctrine fully applies to any stream navigable by any boat, including a recreational craft, to the limit of its navigability. *National Audubon, supra* note 14, 33 Cal. 3d at 435 n.17. Some of the doctrine's "consequences" apply to protect the nonnavigable reaches of such streams. *CalTrout I, supra* note 14, 207 Cal. App. 3d at 630, 631. The doctrine clearly applies to appropriations, even on nonnavigable reaches, which injure the values of navigable waters, such as anadromous fisheries. See *National Audubon, supra* note 14, 33 Cal. 3d at 437, which holds that the doctrine applies to an appropriation that affects a downstream lake.

The complaint alleged that the lower reaches of the Guadalupe, Coyote, and Stevens Rivers, near San Francisco Bay, are navigable. Upstream appropriations on those streams and their tributaries degrade the public-trust values of the navigable reaches, including the populations and distributions of anadromous fisheries. It further alleged that, through inadequate releases, maintenance of fish barriers, and the other causes discussed above, the Santa Clara Valley Water District has harmed the fish and wildlife resources of these streams in violation of the public trust doctrine. Complaint, *supra* note 1, ¶¶ 98–99.

25. Garrett Hardin, *The Tragedy of the Commons*, Science, 162, 1968, at 1243–48.

26. Complaint, *supra* note 1, ¶ 57.

27. *Id.* at ¶¶ 57, 83.

28. *Id.*

29. *Id.* at ¶ 83.

30. *See People of the State of California v. Gold Run Ditch & Mining Company,* 66 Cal. 138, 146–47 (1884).

31. *Id.*

32. *Id.* at 146–47.

33. *Id.* at 149–50.

34. As required by 23 C.C.R. § 822 (2005).

35. Complaint, *supra* note 1, 108–12. The restoration program is a "physical solution" that California law permits as an alternative to abandoning appropriation to protect or restore the public trust. *See Peabody v. Vallejo*, 2 Cal. 2d 351, 383–84 (1935); *see also CalTrout I, supra* note 14, 207 Cal. App. 3d at 626, and State Water Resources Control Board, Decision 1631, *supra* note 14, *available at* http://www.waterrights.ca.gov/hearings/decisions/WRD1631.PDF (last visited Aug. 5, 2005).

36. *Id.*

37. The answer is available online at www.n-h-i.org/Guadalupe_River.html (last visited Aug. 5, 2005). or in hard copy from the SWRCB. *See* supra note 8.

38. *Id.* at ¶ 159.

39. *Id.* at ¶¶ 62–67.

40. *Id.* at ¶¶ 10–11.

41. *Id.* at ¶ 16.

42. *Id.* at ¶ 48.

43. *Id.* at ¶¶ 25–27.

44. *Id.* at ¶ 32.

45. *Id.* at ¶¶ 25–47.

46. *Id.* at ¶¶ 42, 47.

47. *Id.* at ¶¶ 39, 43.

48. *Id.* at ¶ 51.

49. *Id.* at ¶¶ 111–12.

50. *Id.* at ¶ 113.

51. *Id.* at Introduction, ¶ 97.

52. This statute provides that "No permit or license to appropriate water in District 4½ [of CDFG] shall be issued by the State Water Rights Board after September 9, 1953, unless conditioned upon full compliance with Section 5937." Fish and Game Code § (5946). Since Mono Lake is in District 4½, the Mono Lake cases did not actually reach the issue whether section 5937 applies equally to other parts of the state.

53. 23 C.C.R. § 782 (2005). See Answer, *supra* note 37, at ¶¶ 92–95.

54. *Id.* at ¶ 88.

55. *Id.* at ¶¶ 96–98.

56. *Id.* at ¶ 97.

57. *Id.* at ¶ 160, *citing City of Long Beach v. Mansell*, 3 Cal. 3d 462 (1970).

58. See Letter from Brian Hunter, CDFG Region 3 Director, and Stan Williams, SCVWD General Manager, to Natural Heritage Institute (Oct. 21, 1997)

("FAHCE Invitation"), *available at* www.n-h-i/org/Guadalupe_River.html (last visited Aug. 5, 2005).

59. *Id.*

60. This chapter uses the term *stakeholders* rather than *parties* to describe the agencies and private entities participating in the Fisheries and Aquatic Habitat Collaborative Effort negotiations. Technically, with the exceptions of Santa Clara Valley Water District, which holds the water-right licenses, and the Guadalupe-Coyote Resource Conservation District, which was the complainant about uses of those licenses, none of these stakeholders obtained party status. The State Water Resources Control Board stayed the complaint proceeding immediately after SCVWD's answer and before interventions could occur.

61. The eventual settlement includes a restoration budget of $146 million (2003). It may be understood as falling between the worst- and best-case litigation scenarios for the stakeholders. For example, SCVWD would not settle for more than its worst-case scenario, and GCRCD would not settle for less than its corresponding scenario. The monetary value of a litigated result was a more substantial driver of the settlement than the foreseeable expenses of litigation, which probably would not have exceeded $1 million for all stakeholders.

62. Personal communication with Al Gurevich (Feb. 2005).

63. FAHCE Settlement, *supra* note 8, at ¶ 7.

64. Cal. Water Code § 179 (2005).

65. FAHCE Invitation, *supra* note, 58.

66. Personal communication with Al Gurevich (Feb. 2005).

67. *Id.*

68. FAHCE Invitation, *supra* note 58, at 1.

69. Personal communication with Al Gurevich (Feb. 2005).

70. *See* Reorganization Plan No. 4 of 1970, § 1, *codified at* 5 U.S.C. app. 1 (2005).

71. 16 U.S.C. §§ 1531–1544 (2005).

72. Effective October 17, 1997, the National Marine Fisheries Service (NMFS) listed Central California Coast steelhead (*Oncorhynchus mykiss*) as threatened under the Endangered Species Act. *See* 62 Fed. Reg. 43,937 (Aug. 18, 1997). The "evolutionarily significant unit" of Central California Coast steelhead includes coastal California streams (from the Russian River to Aptos Creek) and the San Francisco and San Pablo Bays (including the Guadalupe).

73. *Take* means "to harass, harm, pursue, hunt, shoot, wound, kill, trap, capture, or collect an endangered species, or to attempt to engage in any such conduct." 16 U.S.C. § 1532(19). As defined by rule, *harm* includes significant habitat modification or degradation that results in death or injury to listed species by significantly impairing behavioral patterns such as breeding, feeding, or sheltering. *Harass* includes other actions that create the likelihood of injury to listed species to such an extent as to significantly disrupt normal behavior patterns, which include but are not limited to breeding, feeding, or sheltering. 50 C.F.R. §17.3 (2005).

By rule, the National Marine Fisheries Service has extended the protection against take, applicable by statute to endangered species, to include Central Coast steelhead as a threatened species. *See* 65 Fed. Reg. 42,422 (July 10, 2000). Among other things, this rule describes activities associated with on-stream dams and diversions that are likely to cause harm resulting in take, including "Constructing or maintaining barriers that eliminate or impede a listed species' access to habitat or ability to migrate. . . . Constructing or operating dams or water diversion structures with inadequate fish screens or fish passage facilities in a listed species' habitat. . . . Conducting land-use activities in riparian areas and areas susceptible to mass wasting and surface erosion, which may disturb soil and increase sediment delivered to streams." *See* 65 Fed. Reg. 42,472 (2005).

74. *See* U.S. Fish and Wildlife Service (FWS), *available at* http://www.fws.gov (last visited 2005).

75. *See* Reorganization Plan No. 2 of 1939, § 401, *codified at* 5 U.S.C. app. 1 (2005); Reorganization Plan No. 3 of 1940, § 3, *codified at* 5 U.S.C. app. 1 (2005).

76. 16 U.S.C. §§ 661 *et seq.* (2005).

77. *See supra* note 15. It also is undertaking a Watershed Management Initiative that attempts to integrate the many regulatory laws that have water-quality impacts. http://www.waterboards.ca.gov/sanfranciscobay/watershedmanagement. htm (last visited Aug. 5, 2005).

78. Personal communication with Al Gurevich (Feb. 2005).

79. *Id.*

80. Recycled water that has undergone tertiary treatment may be discharged into a stream pursuant to Cal. Water Code §§ 13556, 13576.

81. Steven E. Ehlmann, *Conflict at the Confluence: The Struggle over Federal Floodplain Management,* 74 N.D.L. Rev. 61, 64–65 (1998).

82. Cal. Water Code § 100 (   ).

83. FAHCE Invitation, *supra* note 58, at 1.

84. *Id.* at 1.

85. *Id.* at 1.

86. *Id.* at 2.

87. Personal communication with Al Gurevich (Feb. 2005).

88. FAHCE Invitation, *supra* note 58, at 2.

89. *Id.* at 2.

90. *Id.* at 2.

91. *See* FAHCE Technical Advisory Committee (TAC), *Investigation to Determine Fish-Habitat Alternatives for the Guadalupe River and Coyote and Stevens Creeks, Santa Clara County* (July 1998), *available at* www.n-h-i.org/ Guadalupe_River.html (last visited) ("Limiting Factors Study").

92. Personal communication with Al Gurevich (Feb. 2005).

93. Limiting Factors Study, *supra* note 91.

94. FAHCE Invitation, *supra* note 58, at 1–2.

95. Personal communication with Al Gurevich (Feb. 2005).

96. *Id.*

97. *Id.*

98. *Id.*

99. Limiting Factors Study, *supra* note 91.

100. *Id.*

101. *Id.* at 27–28.

102. *Id.* at 13.

103. Personal communication with Al Gurevich (Feb. 2005).

104. *Id.*

105. Cal. Code Civ. Proc. § 1152 (2005).

106. Personal communication with Al Gurevich (Feb. 2005).

107. *Id.*

108. *Id.*

109. *Id.*

110. *Id.*

111. *Id.*

112. FAHCE Settlement, *supra* note 8.

113. *Id.* at § 1.1.1, referring to Comprehensive Environmental Response, Compensation, and Liability Act (CERCLA), § 107(f), 42 U.S.C. § 9607(f) (2005). The Santa Clara Valley Water District had potential liability under CER-CLA, even though it had never owned or operated the mine. Its downstream dams captured mercury leachate suspended in the river flow and, through a chemical reaction caused by a low oxygen level in reservoirs in hot weather (known as *methylation*), may have changed the chemical composition of the leachate. CERCLA creates strict liability for any person who owns or operates a facility where a hazardous waste is disposed. 42 U.S.C. § 9607(a)(3) (2005).

114. FAHCE Settlement, *supra* note 8.

115. Cal. Pub. Res. Code §§ 21000 *et seq.* (2005).

116. National Environmental Policy Act, 42 U.S.C. §§ 4321–4347 (2005).

117. *See* 50 C.F.R. § 402.14(i) (2005).

118. FAHCE Settlement, *supra* note at §§ 5.3–5.4.

119. *Id.* at § 4.1.2.

120. *Id.* at §§ 5.3.1, 5.4.1.

121. *Id.* at §§ 4.1.2–4.1.3.

122. *Id.* at § 4.1.3.

123. *Id.* at § 9.1.

124. *Id.* at § 5.6.

125. *Id.* at § 2.2.8.

126. *Id.* at § 4.2.4.

127. *Id.* at § 3.1.

128. *Id.* at § 3.2.

129. *Id.* at §§ 9.3.1–9.3.2.

130. Personal communication with Al Gurevich (Feb. 2005).

131. FAHCE Settlement, *supra* note 8.

132. *Id.* at § 6.2.2.

133. *Id.* at § 7.2.

134. *Id.* at § 7.3(A).

135. *Id.* at § 6.1.

136. *See, e.g., id.* at § 6.6.2.1.2.1 (Guadalupe Creek).

137. *Id.; see also* app. E.

138. *Id.*

139. *Id.* at § 6.2.4.5.

140. *Id.* at § 6.6.

141. *Id.* at § 3.1.2.

142. *Id.* at § 6.6.1.1(D).

143. *See, e.g., id.* at § 6.6.2.1.1.

144. *Id.* at § 6.6.2.2.

145. *Id.* at § 6.6.2.2. The Limiting Factors Study found that fish ladders are infeasible at the storage dams because of their respective heights.

146. *Id.* at § 6.6.2.3.

147. *Id.* at § 6.7.3.

148. *Id.* at §§ 7.1–7.2.

149. *Id.* at § 6.2.4.3.

150. *Id.* at § 6.2.4.4.

151. *Id.* at § 7.3(C).

152. *Id.* at § 7.3(B).

153. *Id.* at § 7.3(D).

154. *Id.* at §§ 8.1.1, 6.7.

155. 1998 FCP Settlement Record Document, *supra* note 8, at 4.

156. *Id.*

157. *Id.*

158. *Id.*

159. *Id.*

160. *See* Cal. Water Code app. § 60–1 n.2 (2005).

161. Ehlmann, *Conflict at the Confluence, supra* note 81.

162. *Id.*

163. *Fact Sheet, supra* note 5.

164. *See* Water Resources Development Act, § 401(b), Pub. L. No. 99-662 (1986), *as amended by* Pub. L. No. 101-101 (1989).

165. As required by Clean Water Act, § 401(a), 33 U.S.C. § 1341(a) (2005).

166. As required by Cal. Water Code § 13260 (2005).

167. San Francisco Regional Water Quality Control Board (SFRWQCB), Waste Discharge Requirements and Water Quality Certification for Guadalupe River Project, Order 01-036 ¶¶ 6–7 (Mar. 2001) ("2001 FCP Certification"), *available at* http://www.waterboards.ca.gov/sanfranciscobay/adoporders.htm (last visited Aug. 5, 2005).

168. *Id.*

169. *Id.*

170. *Fact Sheet, supra* note 5.

171. *Id.*

172. 1998 FCP Settlement Record Document, *supra* note 2, at 9–10.

173. 33 U.S.C. § 1365 (2005).

174. Letter from Richard Roos-Collins, Natural Heritage Institute, to Tony Bennetti, General Counsel, SCVWD, and Annette Kuz, District Counsel, Sacramento District of the Army Corps (May 22, 1996) ("CWA Notice"), available at www.n-h-i.org/Guadalupe_River.html (last visited Aug. 5, 2005).

175. *Id.* at 1.

176. *Id.* at 2–5.

177. *Id.* at 2–6.

178. *Id.* at 7.

179. Personal communication with Al Gurevich (Feb. 2005).

180. *Id.*

181. *Id.*

182. *Id.*

183. Complaint, *supra* note 1.

184. Personal communication with Al Gurevich (Feb. 2005).

185. *Id.*

186. *Id.*

187. 1998 FCP Settlement Record Document, *supra* note 2, at 13–14.

188. *Id.*

189. Personal communication with Al Gurevich (Feb. 2005).

190. 1998 FCP Settlement Record Document, *supra* note 2, at 15–21, at App. B, Ground Rules, at 97 *et seq.*

191. *Id.*

192. *Id.*

193. *Id.*

194. *Id.*

195. *Id.*

196. *Id.* at 37.

197. Dispute Resolution Memorandum regarding Construction, Operation, and Maintenance of the Guadalupe Flood Control Project (July 1998), ("1998 FCP Settlement"), *available at* www.n-h-i.org/Guadalupe_River.html (last visited Aug. 5, 2005).

198. *Id.*

199. *Id.* at § II.C.

200. *Id.* at § IV.1–2.

201. *Id.* at § V.1.

202. Supplement to Dispute Resolution Memorandum regarding Construction, Operation, and Maintenance of the Guadalupe Flood Control Project (Apr. 1999), *available at* www.n-h-i.org/Guadalupe_River.html (last visited Aug. 5, 2005).

203. *See supra* note 101.

204. 2001 FCP Certification, *supra* note 167, at Finding ¶¶ 13, 18.

205. *Id.* at Ordering Provisions ¶ D.3.

206. *Id.; see also* Finding ¶ 20. The Santa Clara Valley Water District retains its legal responsibility for compliance with the certification. The Adaptive Management Program does not create a joint enterprise in that sense.

207. Personal communication with Al Gurevich (Feb. 2005).

208. *Fact Sheet, supra* note 5.

209. Lisa Owens Viani, *Where the River Came Last,* Landscape Architecture, Feb. 2005, at 48.

210. Personal communication with Al Gurevich (Feb. 2005).

211. *Id.*

212. *Id.*

213. *Id.*

214. *Id.*

215. *Id.*

216. *Id.*

217. *Id.*

218. SFRWQCB, Waste Discharge Requirements and Water Quality Certification for Lower Guadalupe River Flood Protection Project, Order R2-2002-0089 (Sept. 2002) ("2002 FCP Certification"), *available at* http://www.waterboards.ca.gov/sanfranciscobay/adoporders.htm (last visited Aug. 5, 2005), SFRWQCB, Waste Discharge Requirements and Water Quality Certification for

Upper Guadalupe River Flood Protection Project, Order R2-2003-0115 (Dec. 2003) ("2003 FCP Certification"), *available at* http://www.waterboards.ca.gov/sanfranciscobay/adoporders.htm (last visited Aug. 5, 2005).

219. *Fact Sheet, supra* note 5.

220. 2002 FCP Certification, *supra* note 218, at Ordering Provisions ¶ D.30, Findings ¶¶ 22–23; 2003 FCP Certification, *supra* note 218, at Ordering Provisions ¶¶ D.29–30.

221. The cost of these measures, while not estimated in the certifications, will probably exceed $50 million.

222. 2002 FCP Certification, *supra* note 218, at Ordering Provisions ¶ D.24, Findings ¶¶ 18, 21; 2003 FCP Certification, *supra* note 218, at Ordering Provisions ¶ D.16, Findings ¶ 16.

223. 2002 FCP Certification, *supra* note 218, at Ordering Provision ¶ D.24, Findings ¶¶ 18, 21; 2003 FCP Certification, *supra* note 218, at Ordering Provisions ¶ D.28.

224. 2002 FCP Certification, *supra* note 218, at Ordering Provisions ¶ D.30, Findings ¶¶ 22–23; 2003 FCP Certification, *supra* note 218, at Ordering Provisions ¶¶ D.29–30.

225. 2002 FCP Certification, *supra* note 218, at Ordering Provisions ¶¶ D.29–30; 2003 FCP Certification, *supra* note 218, at Ordering Provisions ¶ D.32.

226. 2001 FCP Certification, *supra* note 167; 2002 FCP Certification, *supra* note 218; 2003 FCP Certification, *supra* note 218.

227. *Id.*

228. *Id.*

229. *Id.*

7

Bankside Federal

Melissa Samet

A Corps Problem

Both the plight and potential of the urban river are inextricably inter-twined with the United States Army Corps of Engineers. Propelled by the Flood Control Act of 1936, which declared that the federal government should be involved in flood-control activities nationwide, the Army Corps began planning urban river projects in earnest. In far too many cases, the Army Corps opted to encase urban rivers in concrete and con-fine them between levees to speed water past developed areas. Once-vibrant rivers were destroyed along with urban character, green space, healthy waters, and fish and wildlife habitats. And many times, these projects actually increased flooding downstream.

Unfortunately, outdated and destructive approaches to flood protec-tion and other water-resource challenges remain well entrenched at the Army Corps. Until the agency is forced to change its approach, commu-nities will need to fight project by project to ensure that their hometown rivers do not fall prey to the ecological and economic pitfalls that cur-rently pervade Army Corps project planning.

A National Legacy of Ecological Damage to Rivers

The Army Corps wields enormous control over the nation's rivers, coasts, and wetlands. The agency manages the nation's inland waterway system, dredges ports and harbors, and constructs flood-control and restoration projects.[1] The Army Corps has constructed 8,500 miles of levees, more than 600 dams, 11,000 miles of navigation channels, and countless miles of seawalls, jetties, and artificial beaches. It manages more than 380 lakes and reservoirs and maintains more than 920 coastal and inland harbors.[2] Since 1990, when it was given a new environmental protection mission, the Army Corps has embarked on an ever-growing number of restoration projects that often focus on undoing the damage caused by earlier Army Corps projects.[3]

While Army Corps projects have produced certain economic benefits, they have also caused great environmental harm. The Mississippi River is a prime example. The Army Corps' construction and operation of navigation dams and river-control structures, routine dredging, and water-level regulation, have profoundly altered the natural processes of the Mississippi River. Army Corps–built levees have severed the river's connection to much of its historic floodplain and altered the river's natural hydrology, and stone and concrete riprap now line much of the river's banks. These changes have caused significant losses to the river's backwater, side channel, and wetland habitats.[4] They are also a primary cause of the loss of more than 1,900 square miles of Louisiana's coastal wetlands.[5] These losses have in turn caused severe declines in water quality and considerable harm to fish and wildlife. It is not surprising that Army Corps projects have been cited as a major cause of the dramatic decline in North America's freshwater species, which are disappearing five times faster than land-based species.[6]

Flowing past such urban centers as Minneapolis, St. Paul, Davenport, St. Louis, Memphis, Vicksburg, Baton Rouge, and New Orleans, the impacts of the physical alterations to the river cannot be divorced from the urban core. This was made tragically clear when Hurricane Katrina stormed into New Orleans. Vital coastal wetlands lost to the overengineering of the Mississippi River and other Army Corps projects in the region were not available to reduce the size of Hurricane Katrina's storm surge before it reached New Orleans.[7]

The Army Corps' legacy can be seen on many urban rivers across the country. For example, as discussed in chapter 2 in this book, the Army Corps has encased almost the entire Los Angeles River in concrete. The Flint River in Flint Michigan provides another stunning example. In 1966, the Army Corps reduced a mile of the Flint to a bizarrely shaped concrete canal. For good measure, the river's banks were also lined in concrete. As succinctly described by the Army Corps, this project "eliminated the natural ecosystem, making it devoid of flora and fauna."[8] The

Figure 7.1
Flint River flood-control project, Flint, Michigan. Photograph from the archives of the Army Corps of Engineers.

Army Corps is now examining the possibility of restoring the river by removing the 1966 project.[9]

Flawed Army Corps planning continues to produce far too many projects that cause unnecessary harm, fare contrary to modern science, and ignore sound economics.[10] Understanding the flaws in this planning process can provide important lessons for protecting and restoring urban rivers nationwide.

An Agency in Need of Reform

[I] can't think of a government agency in more dire need of reform than the Corps.
—Former Senate Majority Leader Tom Daschle (D-SD)[11]

The way I see it, the Corps is an agency that likes projects, no matter what they do to the environment. Give them a dollar and they'll push it any way you want.
—Representative Jack Kingston (R-GA)[12]

[The Corps'] mistakes led to the deaths of more than 1,000 residents of this metro area.
—Editorial, New Orleans *Times-Picayune*[13]

During the past decade, the National Academy of Sciences, the Government Accountability Office (GAO), federal agencies, and independent experts have all issued studies highlighting substantial problems with the Army Corps' project planning process.[14] Since all Army Corps projects go through the same basic planning process, urban river projects are not immune from problems that include institutional biases, outdated planning guidance, key errors in planning documents, and the failure to mitigate the damage caused by Army Corps projects.

All Corps projects are developed and evaluated using a set of basic rules known as the *Principles and Guidelines* (*P&G*). The *P&G* dictate how the Army Corps will consider environmental impacts, evaluate project costs and benefits, and select project alternatives.[15] While there is specific guidance for calculating the benefits from urban flood damage-reduction projects, the bulk of the Army Corps' analysis of urban river projects is the same as for other types of Army Corps projects.[16] The *P&G* have not been updated since they were written in 1983.[17]

As part of its evaluation, the Army Corps must show that flood-protection and navigation projects will produce a positive dollar return on the taxpayers' investment.[18] The Army Corps does this by preparing a benefit-cost analysis that compares projected project benefits to projected project costs. This analysis goes to the very heart of the Army Corps' ability to recommend a specific project. If the benefits do not exceed the project's costs, the Army Corps may not recommend that project for construction. The inherent difficulties in placing an economic value on restoration efforts led Congress to legislate a positive benefit-cost ratio for such projects.[19]

The Army Corps' recommended project, supporting analyses, and mitigation plans (where projects will cause more than minimal impacts) are presented to Congress in reports known as *feasibility studies*.[20] Congress also receives an environmental review of the project (as required by the National Environmental Policy Act) and any additional evaluations required by other federal laws. These studies guide the Army Corps' ultimate project recommendation—unless a legal challenge or other intervention forces the Army Corps down another path.

Once the Army Corps recommends a project, Congress must decide whether to authorize the Army Corps' plan for construction. A valid decision can be made only if the Army Corps' recommendations are based on studies that are scientifically, economically, and legally sound. Regrettably, the record shows that this is often not the case. As the GAO recently told Congress, "The Corps' track record for providing reliable information that can be used by decision makers to assess the merits of specific Civil Works projects and for managing its appropriations for approved projects is spotty at best."[21]

An Institutional Bias for Large-Scale Structural Projects

The most intractable problem with the Army Corps' planning process may be the agency's overriding institutional bias for recommending large and environmentally damaging structural projects. This problem has been recognized by two National Academy of Sciences panels and the Department of the Army inspector general.[22]

The impact of the Army Corps' bias was aptly summed up by a former regional director of the U.S. Fish and Wildlife Service in 2000: "The Corps still doesn't get it. They still think they can defeat Mother Nature with brilliant engineering. They talk about the environment, but they don't really believe in it."[23]

This bias is driven in large part by the long outdated *P&G*, which encourage large-scale structural solutions over less environmentally harmful, nonstructural approaches. The *P&G* direct the Army Corps to select the National Economic Development (NED) plan unless there are "overriding reasons" for selecting another approach.[24] The NED plan is the one that produces the greatest economic benefit to the nation, consistent with protecting the nation's environment. This relegates environmental protection to a secondary status, as the Army Corps focuses almost exclusively on producing projects that maximize net economic benefits.

The *P&G* also prevent the Army Corps from assigning an adequate value to the full panoply of benefits provided by natural healthy systems, which in turn can "discourage consideration of innovative and nonstructural approaches to water resources planning."[25] For example, the Army Corps frequently fails to assess the true ecological and economic value of maintaining a natural river or wetland system.[26] These values include reducing flooding, cleaning and filtering water supplies, providing vital habitat for fish and wildlife, and supporting vibrant recreational industries.

Under current rules, the Army Corps can—and does—claim economic project benefits from draining wetlands.[27] It is hard to imagine a practice more at odds with the mandates of the Clean Water Act, the National Environmental Policy Act, long-standing national wetlands-protection policies, national farm policies, and national floodplain policies. This practice is equally at odds with the Army Corps' environmental protection mission and its statutorily mandated goal of no net loss of the nation's wetlands.[28]

Two National Academy of Sciences panels have recommended updating the *P&G* "to incorporate contemporary analytical techniques and changes in public values and federal agency programs."[29]

A Pattern of Critical Planning Flaws

Layered on top of the Army Corps' institutional bias is a troubling pattern of critical flaws in Army Corps project analyses.[30] In March 2006, the GAO told Congress that Army Corps studies were so flawed they "did not provide a reasonable basis for decision-making."[31] Army Corps studies reviewed by GAO "were fraught with errors, mistakes, and miscalculations, and used invalid assumptions and outdated data."[32]

For example, in 2003 the GAO found that the Army Corps had overstated the benefits of deepening the main shipping channel in the Delaware River by 200 percent. The Army Corps proposed deepening the channel from forty to forty-five feet from the mouth of the Delaware Bay through Philadelphia Harbor and on to Camden, New Jersey—a distance of more than 100 miles.[33] In fiscal year 2004, the Army Corps estimated that the project would cost $286 million and produce $40.1 million in benefits each year.[34]

But after examining the Army Corps' own data, the GAO could find benefits of only $13.3 million a year, at best. The Army Corps could not explain how it came up with its $40.1 million figure, but it did blame $4.7 million of the differential on a computer error that the agency said "could have occurred when files were transferred from one program to another."[35]

The GAO identified a series of "material errors" in the Army Corps' analysis that included miscalculations, invalid assumptions, and the use of significantly outdated information: "For example, the Corps misapplied commodity growth rate projections, miscalculated trade route distances, and continued to include benefits for some import and export traffic that has declined dramatically over the last decade. In addition, a number of unresolved issues and uncertainties were not factored into the Army Corps' economic analysis, the outcome of which could either increase or decrease the benefits and costs of the project."[36] In the end, the GAO concluded that the Army Corps' economic analysis was so flawed that it could "not provide a reliable basis for deciding whether to proceed with the project."[37]

Deepening 100 miles of river by five feet is not an environmentally benign exercise. Unfortunately, the Army Corps' environmental analysis of the project also contained numerous major flaws. For example, the Army Corps ignored potential water-quality impacts from resuspending toxic sediments during the dredging. It also ignored impacts to human health and air quality caused by "increased diesel emissions from dredging in a three-city area suffering from severe air pollution."[38] The Army Corps also failed to adequately evaluate the damage that the project would cause to water quality, water supplies, and fish and wildlife.[39]

Army Corps analyses often suffer from problems that are far more mundane but no less problematic. In attempting to justify deepening fifty seven miles of the Chesapeake & Delaware Canal from thirty five to forty feet, the Army Corps made "a basic math error that boosted the benefit-cost ratio from a failing 0.65 to a passing 1.21."[40] This error and dozens more were uncovered by four retirees. Media coverage of the math error ultimately forced the Army Corps to suspend the project because it was not economically justified.[41]

Fraud was involved in at least one recent Army Corps analysis. In 2001, the Department of the Army inspector general found that the Army Corps had deceptively and intentionally manipulated data in an attempt to justify what was then estimated to be a $1.2 billion expansion of locks on the Upper Mississippi and Illinois Rivers.[42]

In 2004, this same project failed two reviews by the National Academy of Sciences. The National Academy told the Army Corps that it could not even begin to justify a multibillion national investment in longer locks until it first showed that small-scale traffic-management measures would not achieve the same goal of reducing traffic congestion. The National

Academy also found that the Army Corps' recommendation was based on discredited economic models that produced projections of significantly increased river traffic that could not be reconciled with historic and current traffic levels.[43] The Army Corps ignored these recommendations and just two months later recommended that Congress authorize construction of seven new 1,200-foot locks and implement some small-scale traffic-management measures at a first cost that had escalated to $2.03 billion.[44]

Regrettably, such problems are not new. A 1987 *Washington Post* editorial excoriated the Army Corps' work on the Tennessee-Tombigbee Waterway: "The waterway was justified over the years by egregiously skewed cost-benefit estimates—what you would call lies if your children told them instead of the Corps of Engineers."[45]

Failure to Mitigate Damage Caused by Army Corps Projects

The Army Corps exacerbates the damage caused by its projects by failing to comply with its mitigation mandates. Since 1986, the Army Corps has been required to submit detailed mitigation plans to Congress with every project recommendation, unless the project would cause only minimal harm to the environment.[46] Since 1990, the Army Corps also has had a statutorily mandated "interim goal of no overall net loss of the Nation's remaining wetlands base, as defined by acreage and function, and a long-term goal to increase the quality and quantity of the Nation's wetlands, as defined by acreage and functions."[47]

Despite these legal requirements, the Army Corps has carried out no mitigation at all for the vast majority of its projects. The GAO reports that between 1986 and 2001, only 31 percent of Army Corps projects even had a mitigation plan.[48] Only the most credulous could believe that 69 percent of Army Corps projects produce only minimal harm to the environment, which is the only justification for not implementing mitigation. The Army Corps' own regulations acknowledge that the environment will be harmed by "practically all flood control projects," and even a cursory look at projects with no mitigation plans make it clear that the Army Corps is not complying fully with its mitigation mandates.[49]

For example, the Environmental Protection Agency (EPA) found that the Army Corps' American River Watershed Flood Plain Protection Project would have "adverse environmental impacts that are of sufficient magnitude that EPA believes the proposed action must not proceed as proposed."[50] EPA also found that both the Boston Harbor Navigation Improvements and Berth Dredging Project and the John T. Myers and Greenup Lock Improvements would have "significant environmental impacts that should be avoided in order to adequately protect the environment."[51] Yet the Army Corps did not provide mitigation plans when it asked Congress to approve these projects.[52]

For those few projects where the Army Corps has proposed mitigation, it typically is not carried out in the timeframe required by law.[53] More often than not, Army Corps mitigation also fails to replace habitat destroyed with the same-quality habitat. Instead, the Corps will replace rare aquatic and riparian habitat with fewer acres of more common terrestrial habitat. This out-of-kind mitigation by definition cannot replace lost wetland functions and cannot meet the Army Corps' statutorily mandated goal of no net loss of wetland acres.

For example, while the Army Corps' plan to dredge over 100 miles of the Big Sunflower River in Mississippi will destroy 3,631 acres of wetlands, the Army Corps' mitigation is limited to planting tree seedlings on 1,912 acres of frequently flooded agricultural lands.[54] This is not wetlands mitigation and will not replace the wetland functions that will be lost through the project. Even if this mitigation somehow created 1,912 acres of wetlands, it would still result in a 47 percent loss of wetlands.

The Army Corps also makes little effort to evaluate whether its mitigation efforts are working. For example, as of November 2000, the Army Corps' Vicksburg District had carried out no mitigation monitoring at all on any of its many civil works projects. The Vicksburg District covers portions of three states (Arkansas, Louisiana, and Mississippi).[55]

Flawed Planning has Real Consequences

Flawed planning has real—and sometimes catastrophic—consequences. This was made abundantly clear when Hurricane Katrina hit New Orleans in August 2005.

The devastation of the city was caused in large part by decades of Army Corps water projects that caused the massive destruction of Louisiana's natural storm defenses—its coastal wetlands. The coupe de grace was the Army Corps' design of faulty levees and floodwalls that failed catastrophically in a storm they were supposed to withstand.

Army Corps flood-control and navigation works on the Mississippi River are a primary cause of the loss of some 1,900 square miles of Louisiana's coastal wetlands that would have helped reduce the size of Hurricane Katrina's storm surge before it reached the New Orleans area.[56] Since every 2.7 miles of coastal wetlands can reduce storm surge by one foot, these losses played an undeniable role in the city's destruction.[57] Wetlands lost to Army Corps projects also increased the risk of levee failures in the city. Levees in and near New Orleans "with a buffer of wetlands had a much higher survival rate than those that stood naked against Katrina's assault."[58]

Another Army Corps project, a little-used navigation channel known as the Mississippi River Gulf Outlet, caused more than 20,000 acres of coastal wetland losses and greatly exacerbated the hurricane's impacts by funneling and intensifying Katrina's storm surge directly into New Orleans.[59] The flooding that overwhelmed St. Bernard Parish and the lower Ninth Ward of New Orleans came from the Outlet. The impacts were devastating. Only fifty two of the 28,000 structures in St. Bernard Parish escaped unscathed from Katrina. Scientists and activists had warned the Army Corps for decades that the Outlet was a hurricane highway pointed straight into the heart of the city. For years, calls to close the Outlet had gone unheeded, including a 1998 plea by the St. Bernard Parish Council.[60]

The Army Corps also set the stage for the Katrina disaster when it planned the city's hurricane projection project after Hurricane Betsy hit New Orleans in 1965. Instead of reinforcing levees located at the city's edge, the Army Corps planned an elaborate new system stretching miles into uninhabited wetlands. The Army Corps used the improved property values from the wetlands drained by the project to justify the project's significant cost (estimated in 1978 at $409 million). Many of the wetlands that were developed as a result of the project became the eastern Orleans Parish neighborhoods that suffered the brunt of Katrina's flooding.[61]

Design flaws in the floodwalls and levees were the final straw in the city's destruction. They failed catastrophically in what has been described as "the greatest engineering failure in American history, measured by lives lost, people displaced and property destroyed."[62]

Hurricane Katrina was no more than a category 3 storm by the time it reached New Orleans, a storm event that the floodwalls were supposed to protect against. But Ivor Van Heerden, deputy director of the Louisiana State University Hurricane Center and director of the Center for the Study of Public Health Impacts of Hurricanes in Baton Rouge, has said that the hurricane protection system designed by the Army Corps "wasn't even capable of withstanding a Category One hurricane."[63] The floodwall design did not meet the Army Corps' own guidelines, and the Army Corps knew that the floodwalls were being built on extremely unstable soils that likely warranted a much stronger design.[64]

As disturbingly, the Army Corps ignored crucial data on the need to increase the levee heights. The Army Corps was told as early as 1972 that new weather data showed that the levees needed to be higher than planned to protect New Orleans from stronger hurricanes. This information was not incorporated into the hurricane system's design specifications even though construction did not begin until the 1980s.[65]

The Army Corps acknowledged that "a 'design failure' led to the breach of the 17th Street Canal levee that flooded much of the city during Hurricane Katrina."[66] Independent engineers investigating the levee failures have pointed out design flaws at many other locations, and the American Society of Civil Engineers has said the catastrophic failure is undeniable proof that the Army Corps' design contained "fundamental flaws."[67]

Reform Legislation

Projects developed through flawed planning damage the nation's rivers and wetlands, squander agency resources and tax dollars, and adversely impact recreation, tourism, and other businesses that rely on healthy rivers. As seen in New Orleans, poorly designed projects also put people at risk. When tax dollars and Corps resources are spent on planning, defending, or constructing such projects, the Army Corps has less ability to carry out environmental protection and restoration projects or more deserving flood-control and navigation projects.

A growing national movement (including the Army Corps Reform Network, which currently has more than 140 national, regional, and local member organizations) has been pushing Congress to institute much needed reforms that include updating the *P&G* and the planning regulations that flow from them; requiring fully independent review of Army Corps projects costing more than $25 million or that are deemed to be controversial based on a clear set of criteria; tightening the mitigation requirements for Army Corps projects to ensure that the Army Corps proposes, implements, and monitors mitigation to fully offset the environmental damage caused by its projects; and focusing Army Corps projects on national priorities. Legislation that includes these and other needed reforms has been introduced during each of last three Congresses.[68] It appears that the Army Corps' role in the devastating flooding of New Orleans may finally force Congress to act on these reforms.

Ill-Conceived Projects in the Pipeline

Simply put, the problems at the Army Corps result in bad projects for the nation. This is particularly troubling as the agency is constructing more than 1,500 projects nationwide and has a backlog of projects that would cost some $58 billion to complete.[69] And Congress continues to authorize new projects on a regular basis. In 2000, Congress authorized 195 new Army Corps projects in the Water Resources Development Act.[70]

Some key planning flaws in two major urban river projects are outlined below, but these problems are by no means isolated incidents. Significant flaws have been uncovered in virtually every project reviewed by experts fully independent of the Army Corps.

The Sacramento and American Rivers Confluence

Located at the confluence of the Sacramento and American Rivers, Sacramento, California, has faced the threat of major flooding since its founding.[71] To reduce these flood risks, the Army Corps twice recommended constructing a new dam on the American River near Auburn, California. But the high costs of the proposed dam and its adverse environmental impacts caused Congress to reject each of those proposals.[72]

The strength of the opposition to the proposed Auburn Dam led the Army Corps to recommend a smaller-scale project in June of 1996.[73] This plan involved improving sections of the American and Sacramento Rivers levees, primarily by constructing cut-off walls in the center of the levees to make them more impervious to water seepage which is a major cause of levee failure.[74] These levees protect downtown Sacramento and a largely agricultural area just north of downtown that is being rapidly developed.[75]

It quickly became clear, however, that the Army Corps' levee plans would not provide the promised level of flood protection. As a 2003 GAO report noted: "A severe storm in January 1997 demonstrated vulnerabilities in the American River levees and alerted the Army Corps of the need to do additional work to close the gaps in the cut-off walls at bridges and other areas and extend the depth of some cut-off walls from about 20 feet to about 60 feet."[76]

The considerable design changes needed to address these planning flaws dramatically increased the project's cost. In 1996, the Army Corps estimated that the project would cost $57 million. Just six years later, costs had ballooned to between $270 million to $370 million.[77] These remarkable increases have a very real impact on the project's local sponsors, who are required to pay a percentage of the total project cost.

The flood-protection claims in the Army Corps' original plan were also incorrect. The GAO found that the Army Corps had significantly overestimated both the amount of land and the number of homes that would be protected by the project. "The actual number of protected residential properties was about 20 percent less than the number that the Army Corps estimated."[78] The Army Corps also did not properly assess the value of the properties it did count.[79]

According to the GAO, these mistakes likely ensured that the original project could be recommended by the Army Corps.[80] The project was just barely justified with a 1.1 to 1 benefit-cost ratio when the Army Corps first recommended the plan to Congress.[81]

The Trinity River, Dallas, Texas

The Trinity River flows past Fort Worth and Dallas, Texas, before flowing south to the Gulf of Mexico. Through much of Dallas, the Trinity is surrounded by a unique and vibrant urban oasis, the 5,900-acre Great Trinity Forest.[82]

In 1965, Congress authorized the Army Corps' plan to channelize more than nineteen miles of the Trinity River and its tributaries and construct twenty-two miles of new levees. Planning for this project, known as the Dallas Floodway Extension, was stopped in 1985 when the city of Dallas was unable to convince voters to support a bond to cover the local cost share for the project.[83]

But it is hard to stop a Army Corps project once it is authorized, and the project was resurrected in 1989 after severe flooding hit the Dallas area. Ten years later, the Army Corps finalized a new plan that would extend the existing levees that protect downtown Dallas by more than five miles to protect residential neighborhoods downstream, reroute 3,000 feet of the Trinity River, and cut a 600-foot wide, 3.7 mile long swale through the Great Trinity Forest.[84] More than 30,000 trees will be cut down in the process.[85] Between December 1999 and April 2003, the estimated cost of the project swelled by $27 million to $154.4 million.[86] Though the Army Corps was forced to take another look at the plan by both the White House Office of Management and Budget (OMB) and the U.S. District Court for the Northern District of Texas, the Army Corps did not change its plan in any way.[87]

Once the Dallas Floodway Extension is complete, the city of Dallas, the North Texas Tollway Authority, and the Texas Department of Transportation intend to construct ten lanes of toll road within the Trinity River's floodplain. Most of the roadway will be constructed using material excavated from the swales to be cut in the Trinity Forest during construction of the Dallas Floodway and related projects. Those sections of the roadway that cannot be raised above the 100-year floodplain will be protected by 100-year floodwalls that will further impact the floodplain. A final Record of Decision on the toll road is expected in 2007. Construction could be completed as early as 2010.[88]

Cutting 30,000 trees and gouging out a 3.7-mile long and 600-foot wide swale in the Great Trinity Forest will cause undeniable and significant damage to this valuable urban resource. Realigning the Trinity's channel will destroy much of the reach's instream habitat, and water rushing out of the floodway will increase erosion and siltation. As acknowledged by the Army Corps, the new levees also will "facilitate potential future commercial or residential development,"[89] despite the very real possibility of continued future flooding in the area. The new tollway will add an additional layer of impacts, congestion, and pollution.

The OMB has determined that the Army Corps did not comply with the *P&G* in developing its plan for the Dallas Floodway Extension. OMB also believes that a "fundamentally different project appears to be in order."[90] OMB argued that the Army Corps could achieve the vast majority of project benefits—those attributable to increased flood protection to downtown Dallas—at a fraction of the cost simply by raising the existing downtown levees. Instead of pursuing that option, the Army Corps used the benefits that could have been obtained by that smaller project to justify the much more costly and destructive 1999 plan.[91]

According to a 2004 report by the National Wildlife Federation and Taxpayers for Common Sense: "A less expensive alternative is for the Army Corps to raise the existing Dallas levee, and conduct a voluntary buyout in floodprone neighborhoods, such as Cadillac Heights, which is suffering from toxic contamination. This would provide the most effective flood protection for the Dallas area with dramatically less impact to the floodplain environment."[92] A 2003 report by American Rivers notes that residents of Cadillac Heights, a largely minority neighborhood, "have indicated that their preferred solution to periodic flooding is a voluntary buyout rather than new levees and freeways on their doorstep."[93]

Community Planning Can Help Avert Bad Projects: Napa River, California

Determined community involvement can turn Army Corps projects around. The best example of this may be the Napa River project in Napa, California. After rejecting Army Corps plans to pave and channelize the Napa River three times, the community developed its own plan and convinced the Corps to adopt it.

Located along the Napa River in an area highly susceptible to flooding, the town of Napa has flooded twenty seven times between 1862 and 1997. The most damaging flood occurred in February 1986, when three people died, 2,750 homes were damaged, and 5,000 people were forced to evacuate. Between 1961 and 1997, property damage from flooding in Napa County reached $542 million.[94]

In 1965, Congress authorized a flood-protection project for the Napa River.[95] Ten years later, in 1975, the Army Corps finally presented a plan to the public. The Army Corps proposed its traditional approach to flood control—enlarging the river channel and constraining it with levees. But the town would have none of it. Voters rejected the Army Corps' plan in two separate referendums, and the Army Corps finally gave up and put the project on inactive status.[96]

In October 1988, the project was reactivated in response to severe flooding that had occurred two years earlier. This time it took the Army Corps almost seven years to produce a "new" plan for the Napa River. Released in April 1995, this new plan followed the same flawed approach rejected by the voters almost twenty years earlier: "The plan's traditional approach—enlarging the River channel and constraining the river within that channel—was met with an underwhelming response in Napa. The proposal was seen to be 'environmentally insensitive' at best, and did not inspire aesthetically. Lacking local support, the new plan appeared to be dead on arrival."[97]

Rather than give up on a project altogether, the community created a "precedent-setting cooperative coalition" of citizens, environmental organizations, and business groups ranging from the Friends of the Napa River and the Sierra Club to the local Farm Bureau and the Chamber of Commerce.[98] The community also hired consultants and worked closely with the Army Corps and other federal and state agencies.[99]

After thousands of hours of meetings spread out over a year and a half, the coalition reached consensus on a "living river" design for the Napa River. The design was so popular that less than one year later, the county's voters agreed to a half-cent sales tax increase to fund the local cost share for the project. The community remains an active participant in the continued planning and implementation of the project, and has established both a Technical Advisory Panel and a Financial Oversight Committee.[100]

The project seeks to work with the river's natural processes. Portions of the river are being reconnected to its historic floodplain, and over 600 acres of tidal wetlands will be restored. The project also creates a unique dry bypass area so that the river can follow its traditional high-water path without flooding homes and businesses. Other elements of the project include removing and raising bridges that block the flow of high water, cleaning up contaminated areas, building riverside trails and promenades, and monitoring impacts to fish populations.[101]

The project is held up by many—including the Army Corps—as one of the most effective approaches to urban river planning. For the Army Corps, however, Napa remains the exception to the rule.

Lessons for Urban Rivers

Important lessons can be drawn from the problems associated with these and other Army Corps projects. Applying these lessons to new urban river projects may improve the future of urban rivers nationwide.

Lesson 1: Recognize the Problem

The Army Corps' planning process is plagued by critical problems that lead to ill-conceived and costly projects that threaten the ecological health of urban rivers. Ultimately, these problems will be solved only through passage of comprehensive legislative reforms.

Until such reforms become law that is strictly complied with, communities seeking help with urban river problems should proceed cautiously in seeking the Army Corps' involvement. This is true for flood-damage reduction, riverfront revitalization, and urban river-restoration projects. Where the Army Corps is involved, the state, local governments, community groups, conservation and taxpayer organizations, and citizens should carefully consider taking some or all of the additional steps discussed below.

Lesson 2: Bring Good Projects to the Army Corps

Communities seeking the Army Corps' assistance with urban river projects should strongly consider developing their own plan to bring to the Corps. Communities may want to hire outside planners and consultants to assist in developing a plan that is ecologically sound and that meets the goals of a healthy, accessible, and vital riverfront. This upfront investment can pay great dividends down the road. Communities will then need to work to ensure that the community's preferred plan is implemented. The Napa River project, which is highlighted by the Army Corps as one of its most successful urban river-restoration efforts, was developed in this way.

Community members should not leave planning or planning oversight responsibilities in the hands of the local project sponsor. Many environmentally destructive projects have been implemented with the full support of the local project sponsor, and even ecologically sophisticated local sponsors can be enticed to support a bad project.

When developing and commenting on urban river plans, communities should keep in mind some of the conflicts inherent in urban river revitalization and ecological restoration (that is, the restoration of natural river processes to create a self-sustaining healthy ecosystem). While single-focus restoration plans may be difficult to implement on some urban rivers, communities should work to maximize the ecological benefits to be gained from every urban river project and should make reestablishing the ecological health of the river a primary goal wherever possible. Communities also should consider whether ecological restoration can be the single goal for at least certain segments of urban rivers.

Communities may want to utilize the 2004 guidebook *Ecological Riverfront Design: Restoring Rivers, Connecting Communities*, prepared by American Rivers and the American Planning Association. This guidebook provides an extensive evaluation of ecological riverfront planning and design and identifies key principles and design guidelines that could be extremely useful as communities develop and comment on urban river plans.[102]

These design principles recommend that projects (1) "Preserve natural river features and functions," (2) "Buffer sensitive natural areas," (3) "Restore riparian and in-stream habitats," (4) "Use nonstructural alternatives to manage water resources," (5) "Reduce hardscapes," (6) "Manage stormwater on site and use nonstructural approaches," (7) "Balance recreational and public access goals with river protection," and (8) "Incorporate information about a river's natural resources and cultural history into the design of riverfront features, public art, and interpretive signs."[103]

Communities also should consider the following criteria for successful river restoration identified in 2005 by a group of twenty-two preeminent river scientists and planners: (1) "the design of an ecological river restoration project should be based on a specified guiding image of a more dynamic, healthy river that could exist at the site," (2) "the river's ecological condition must be measurably improved," (3) "the river system must be more self-sustaining and resilient to external perturbations so that only minimal follow-up maintenance is needed," (4) "during the construction phase, no lasting harm should be inflicted on the ecosystem," and (5) "both pre- and post-assessment must be completed and data made publicly available."[104]

Lesson 3: Engage State and Other Federal Agencies
States and other federal agencies can play a key role in Army Corps project planning and, in some instances, can stop particularly ill-conceived projects altogether.

The United States Environmental Protection Agency (EPA) and the United States Fish and Wildlife Service (FWS) are required by law to review and comment on Corps projects. EPA must comment on the Army Corps' environmental review under the National Environmental Policy Act and ultimately is responsible for ensuring that Corps projects comply with the Clean Water Act. FWS must review and comment on Corps projects under the Fish and Wildlife Coordination Act and must work with the Army Corps to ensure that projects comply with the federal Endangered Species Act. The National Oceanic and Atmospheric Administration (NOAA) Fisheries shares these duties when projects have potential impacts on marine species.

The states can significantly restrict and, in some instances, can effectively veto Army Corps projects that do not meet state water-quality standards under section 401 of the Clean Water Act. State fish and wildlife agencies also have the ability to provide detailed recommendations on needed mitigation through the Fish and Wildlife Coordination Act. And of course, state agencies can participate in the National Environmental Policy Act review and in any state-required environmental reviews.

Despite these legal authorities, federal and state agencies often lack the resources and the political will to effectively fight destructive Army Corps projects. As a result, it is important to work closely with agency staff to make sure they understand the full implications of a Corps plan. It is also important to provide vocal public support for agencies willing to work aggressively to improve Army Corps projects.

Lesson 4: Promote Outside-Project Review

Every Army Corps study should be fully and carefully reviewed by federal and state resource agencies, independent experts (including legal experts), and the public. Voluminous planning documents and significant study costs should not lull reviewers into believing that a Army Corps study is necessarily technically sound, based on the most current science and data, or legally sufficient.

Reviewers should carefully analyze all key assumptions, models, and data and should examine the underlying studies and documents relied on by the Army Corps whenever possible. Reviewers also should take special care to identify key issues that have not been addressed in a Army Corps study. For example, Army Corps environmental reviews often ignore impacts on entire classes of species (amphibians are a prime example). The Army Corps also has a penchant for ignoring cumulative impacts, impacts from resuspending and disposing of toxic sediments, project promotion of increased flooding downstream, and a host of other potentially significant impacts. Experts may be needed to analyze key elements of Army Corps studies, including particularly the Army Corps' hydrologic analysis and benefits calculations.

Because of the voluminous nature of Army Corps studies, experts and citizens may choose to analyze only certain key analyses. Such reviews can be extremely valuable and should be encouraged. However, if possible, reviewers should strive to review the study as a whole to see where flaws in one analysis can cause problems in others. For example, where a Army Corps study underestimates a project's adverse impacts to wetlands, the needed mitigation will also be underestimated. This in turn will produce a faulty cost estimate, since the cost of mitigation is a project cost.

Citizens should not be deterred if they lack experience in reviewing Army Corps project studies or if they are greeted with hostility or ridicule while attempting to obtain answers to important questions. Citizen review has uncovered substantial problems with Army Corps projects in the past, and citizen review undoubtedly will continue to do so in the future. Moreover, in many circumstances, citizens will be the only source of knowledge on specific community needs, local problems, and potential local impacts.

Lesson 5: Consider the Full Range of Options to Improve Projects
Ultimately, it may not be possible to stop or improve a destructive Army Corps project without outside help. In these instances, Congress, the media, and the courts can be valuable sources of assistance.

Members of the United States Congress play a key role in every Army Corps project. They can obtain or prevent authorization of a project and obtain or prevent funding for a project. Members of Congress also can request GAO and National Academy of Sciences studies to evaluate specific Army Corps projects. Communities and citizens should discuss key problems with Army Corps studies with their members of Congress early and often. All too often, these key decisionmakers will have heard only from project proponents and will have few insights into problems with particular projects.

Media coverage of key planning problems and potential project impacts also can be extremely useful in highlighting the need for project improvements. Coverage of key flaws or public opposition can also be very useful in convincing members of Congress to pay careful attention as the planning process proceeds.

Where legal challenges are justified, they can be used to stop, redirect, or delay project implementation.

Lesson 6: Working without the Army Corps
Despite the financial and technical resources that the Army Corps can bring to bear on urban river projects, the agency may not always be the best partner. As discussed throughout this essay, Army Corps planning often produces projects that are far less than ideal—and in some instances produces projects that simply should never be built.

Unfortunately, regardless of how destructive an Army Corps project may be, it can be extremely difficult to stop if it gains the support of a powerful constituency. On the flip side, it can be equally difficult to actually construct a project planned by the Army Corps, even if that project is ecologically sound and fully supported by the local community. This is because many hundreds of Army Corps projects can by vying for a limited amount of funding at any given time.

The difficulty in obtaining construction funding is perhaps best exemplified by the Army Corps' enormous construction backlog. It is estimated that it would cost $58 billion to construct all the Army Corps projects already authorized for construction. At the current rate of construction funding for the agency (roughly $1.8 billion annually, not including emergency spending), it would take more than thirty-two years to complete all the projects on the Army Corps' books, assuming that not a single new project was authorized during that time.[105]

Another drawback to working with the Army Corps is that Army Corps projects often are extremely costly. Thus, while Army Corps projects ensure a significant amount of federal funding, the local cost-share requirements can be overwhelming. For some smaller-scale projects, it may be less expensive for a city or town to construct a project on its own than to pay the local cost share for the Army Corps' proposed plan.[106] In addition, actual Army Corps project costs often outstrip original estimates at an astonishing pace. This can place very real financial hardships on local project sponsors who must pay a set percentage of the total—and not the estimated—project costs.

Conclusion

The United States Army Corps of Engineers wields enormous control over the nation's rivers, streams, and wetlands. Unfortunately, that power has often been ill used, and the tragic results can be seen on urban rivers across the country. Until Congress makes fundamental changes to the rules guiding the Army Corps, it is incumbent on the public, local governments, the states, and other federal agencies to make sure that the Army Corps does not build projects that cause more harm than good. The lessons presented in this chapter provide a starting point for undertaking that difficult but important task.

Notes

1. The Army Corps of Engineers also manages the Clean Water Act's section 404 program, through which it has the authority—but typically not the political will—to significantly limit damage from private activities to virtually every river, stream, and wetland in the nation. 33 U.S.C. § 1344, Clean Water Act § 404 (2005). The Corps shares this authority with the U.S. Environmental Protection Agency. *Id.* The Corps rarely denies section 404 permit requests. In 2001 and 2002, the Corps denied fewer than 1 percent of permits requested. U.S. Army Corps of Engineers, *Wetland Mitigation Report Card First Steps*, Presentation by Mark Sudol, Chief of the Regulatory Program to the Stakeholders Meetings, National Wetlands Mitigation Action Plan, July 30, 2003, *available at* http://www2.eli.org/pdf/mitigation_forum_2003/sudol_presentation2.pdf (last visited June 1, 2005). The *St. Petersburg Times* reports that "in 2003 the Corps approved more than 3,400 wetlands permits in Florida, more than any other state. It denied none." Craig Pittman and Matthew Waite, *They Won't Say No*, St. Petersburg Times, May 22, 2005, *available at* http://www.sptimes.com/2005/05/22/State/They_won_t_say_no.shtml (last visited May 28, 2005).

2. U.S. Army Corps of Engineers, Information Paper, *Civil Works Program Statistics*, Feb. 17, 2005 ("*Civil Works Program Statistics Fact Sheet*"), *available at* www.usace.army.mil/inet/functions/cw/hot_topics/didyouknow.pdf (last visited May 28, 2005).

3. The Water Resources Development Act of 1990 established a new mission for the Corps: "The Secretary shall include environmental protection as one of the primary missions of the Corps of Engineers in planning, designing, constructing, operating, and maintaining water resources projects." 33 U.S.C. § 2316(a) (1990). As of January 2004, the Corps was constructing eighty one specifically authorized restoration projects, and 19 percent of the Corps' total appropriation ($866.6 million) went toward restoration projects in fiscal year 2004. The largest of these projects include the Comprehensive Everglades Restoration Program, Columbia River Fish and Wildlife Mitigation, Upper Mississippi River Environmental Management

Program, and the Missouri River Fish and Wildlife Mitigation. *Civil Works Program Statistics Fact Sheet, supra* note 2. Controversy surrounds the Corps' efforts in many of these projects, including, notably, the Everglades Restoration Program. *See* Michael Grunwald, *The Swamp* (2006). This is perhaps not surprising since restoration projects are diametrically opposed to traditional civil works projects. As the Institute for Water Resources points out, "Whereas traditional civil works projects generally rely on management measures to eliminate hydrologic extremes, ecosystem restoration generally requires measures to re-establish natural hydrologic variability." Institute for Water Resources, *Policy Studies Program, Improving Environmental Benefits Analysis in Ecosystem Restoration Planning,* IWR Report 03-PS-3, at xiv, 11 (October 2003), *available at* http://www.iwr.usace.army.mil/iwr/pdf/envirobenefits.pdf (last visited May 28, 2005).

4. *E.g.*, U.S. Geological Survey, *Ecological Status and Trends of the Upper Mississippi River System 1998: A Report of the Long-Term Resource Monitoring Program* 16-1 to 16-12 (April 1999); Rock Island District, U.S. Army Corps of Engineers, *Report to Congress, An Evaluation of the Upper Mississippi River System Environmental Management Program* 2–3 (December 1997).

5. *E.g.*, U.S. Geological Survey, News Release, *Without Restoration, Coastal Land Loss to Continue* (May 21, 2003), *available at* http://www.nwrc.usgs.gov/releases/pr03_004.htm (last visited May 22, 2006).

6. Anthony Ricciardi & Joseph B. Rasmussen, *Extinction Rates of North American Freshwater Fauna*, Conservation Biology, Oct. 1999, at 1220.

7. *E.g.*, Bob Sullivan, *Wetlands Erosion Raises Hurricane Risks, Natural Storm "Speed Bump" around New Orleans Now Missing*, MSNBC, Aug. 21, 2005, *available at* http://www.msnbc.msn.com/id/9118570/ (last visited May 23, 2006); Bob Marshall, *Studies Abound on Why the Levees Failed. But Researchers Point Out That Some Levees Held Fast Because Wetlands Worked as Buffers during Katrina's Storm Surge*, New Orleans Times Picayune, Mar. 23, 2006, *available at* http://www.nola.com/search/index.ssf?/base/news-3/1143101527153500.xml?nola (last visited Mar. 27, 2006).

8. U.S. Army Corps of Engineers, *Project Fact Sheet: Flint River and Swartz Creek Project, Flint, Michigan* (Feb. 24, 2004) (*"Flint River Fact Sheet"*), *available at* http://www.lre.usace.army.mil/_kd/Items/actions.cfm?action=Show&item_id=37 37&destination=ShowItem (last visited May 28, 2005); *see also* J. M. Leonardi & W. J. Gruhn, *Flint River Assessment*, Special Report 27, Michigan Department of Natural Resources, Fisheries Division, Ann Arbor, Michigan (2001), at 29.

9. *Flint River Fact Sheet, supra* note 8. The Corps is considering the restoration work under its section 1135 programmatic authority, which allows the Corps to plan and construct projects to restore habitat damaged by past Corps projects. 33 U.S.C. § 2309a (2005). Section 1135 refers to the section of the Water Resources Development Act of 1986 that established this programmatic authority for the Corps.

10. The Corps has a different perspective on the history of its water-project planning. *See, e.g.*, Martin Reuss, *Designing the Bayous: The Control of Water in the Atchafalaya Basin 1800–1995* (2004). Reuss is a senior historian for the Corps of

Engineers, and this book was originally published by the U.S. Army Corps of Engineers in 1998.

11. Michael Grunwald, *Working to Please Hill Commanders*, Washington Post, Sept. 11, 2000, at A1.

12. Michael Grunwald, *In Everglades, a Chance for Redemption*, Washington Post, Sept. 14, 2000, at A1.

13. Editorial, *Divided We Flood*, New Orleans Times Picayune, Feb. 8, 2006.

14. *E.g.*, U.S. Government Accountability Office, *Corps of Engineers, Observations on Planning and Project Management Processes for the Civil Works Program* GAO-06-529T (March 2006) (finding that recent Corps studies were so flawed they could not provide a reasonable basis for decision-making and that the problems at the agency were systemic in nature and prevalent throughout the Corps' civil-works portfolio); American Society of Civil Engineers, *External Review Panel Progress: Report Number 1* (Feb. 2006) (finding that the catastrophic failure of the Corps' New Orleans hurricane-protection system demonstrated that the project design and development was fundamentally flawed); R. B. Seed, P. G. Nicholson, et al. *Preliminary Report on the Performance of the New Orleans Levee Systems in Hurricane Katrina on August 29, 2005*, Report No. UCB/CITRIS – 05/01 (Nov. 2005) (finding that that three major breaches in New Orleans levee systems appear to have resulted from failings in the Corps' design and construction oversight and that many of the other levees and floodwalls that failed due to overtopping might have performed better if conceptually simple details had been added or altered during their original design and construction); U.S. Government Accountability Office, *Army Corps of Engineers: Improved Planning and Financial Management Should Replace Reliance on Reprogramming Actions to Manage Project Funds*, GAO-05-946 (Sept. 2005) (finding that the Corps' excessive use of reprogramming funds is being used as a substitute for an effective priority-setting system for the civil-works program and as a substitute for sound fiscal and project management); National Academy of Sciences, National Research Council, *Review of the U.S. Army Corps of Engineers Restructured Upper Mississippi River–Illinois Waterway Feasibility Study: Second Report* (Oct. 2004) (finding that flaws in the Corps' models used to predict demand for barge transportation preclude a demonstration that expanding the locks is economically justified and that the Corps has not given sufficient consideration to inexpensive, nonstructural navigation improvements that could ease congestion at the existing locks); National Academy of Sciences, National Research Council, *Review of the U.S. Army Corps of Engineers Restructured Upper Mississippi River–Illinois Waterway Feasibility Study* (First Report, 2004) (finding that the Corps' models are fundamentally flawed, traffic projections are significantly overstated, and the Corps cannot assess the need for the project until small-scale measures to reduce congestion are implemented); U.S. Commission on Ocean Policy, *An Ocean Blueprint for the Twenty-first Century: Final Report of the U.S. Commission on Ocean Policy* (Sept. 2004) (recommending, among other things, that the National Ocean Council review and recommend changes to the Corps' civil-works program to ensure valid, peer-reviewed cost-benefit

analyses of coastal projects, provide greater transparency to the public, enforce requirements for mitigating the impacts of coastal projects, and coordinate such projects with broader coastal planning efforts); Congressional Research Service, *Agriculture as a Source of Barge Demand on the Upper Mississippi and Illinois Rivers: Background and Issues*, RL32401 (May 2004) (finding that grain traffic forecasts being used by the Corps to justify lock expansion on the Upper Mississippi River were overly optimistic); National Academy of Sciences, National Research Council, *U.S. Army Corps of Engineers Water Resources Planning: A New Opportunity for Service* (2004) (recommending modernizing the Corps' authorities, planning approaches, and guidelines to better match contemporary water-resources management challenges); National Academy of Sciences, National Research Council, *Adaptive Management for Water Resources Project Planning* (2004) (recommending reforms to ensure effective use of adaptive management by the Corps for its civil-works projects); National Academy of Sciences, National Research Council, *River Basins and Coastal Systems Planning within the U.S. Army Corps of Engineers* (2004) (recommending needed changes to the Corps' current planning practices to address the challenges of water-resources planning at the scale of river basins and coastal systems); National Academy of Sciences, National Research Council, *Analytical Methods and Approaches for Water Resources Planning* (2004) (recommending needed changes to the Corps' *Principles and Guidelines* and planning guidance policies); U.S. General Accounting Office, *Improved Analysis of Costs and Benefits Needed for Sacramento Flood Protection Project*, GAO-04-30 (Oct. 2003) (finding that the Corps dramatically miscalculated the costs and benefits of the Sacramento Flood Control Project in California, overcounted the residential properties that would be protected, miscalculated the area that would be protected, and used an inappropriate methodology to calculate prevented flood damages and recommending that the Corps improve its cost-benefit analysis and cost-accounting procedures); Pennsylvania Transportation Institute (PTI), *Analysis of the Great Lakes/St. Lawrence River Navigation System's Role in U.S. Ocean Container Trade* (Aug. 2003) (finding fundamental flaws in the Corps' plan to expand the Great Lakes navigation system, including a number of factors not considered by the Corps that make the Great Lakes ports unattractive to international containerized cargo); Pew Oceans Commission, *America's Living Oceans, Charting a Course for Sea Change, A Report to the Nation, Recommendations for a New Ocean Policy* (May 2003) (recommending enactment of "substantial reforms" of the Corps, including legislation to ensure that Corps projects are environmentally and economically sound and reflect national priorities); U.S. General Accounting Office, *Oregon Inlet Jetty Project: Environmental and Economic Concerns Need to Be Resolved*, GAO-02-803 (Sept. 2002) (finding that the Corps' economic analysis did not provide a reliable basis for decision-making because it relied on outdated and incomplete data and unsupported assumptions, failed to account for risk and uncertainty in key variables that could significantly affect the project's benefits and costs, and failed to propose adequate mitigation for the environmental impacts of the project); U.S. General Accounting Office, *Delaware River Deepening Project: Comprehensive Reanalysis Needed*, GAO-02-604 (June 2002) (finding that the Corps' analysis did not provide a reasonable basis for decision-making because it relied on a flawed benefit-cost analysis that was based on invalid assumptions and outdated information); U.S. General Accounting Office, *Scientific Panel's Assessment of Fish and Wildlife Mitigation Guidance*, GAO-02-574 (May 2002) (reporting that the Corps has proposed no mitigation for 69 percent of projects constructed since 1986 and that for those few projects where the Corps does carry out mitigation, it was doing so under the timelines required by law only 20 percent of the time); National Academy of Sciences, National Research Council, *Review Procedures for Water Resources Planning* (2002) (recommending creation of a formalized process to independently review costly or controversial Corps projects); National Academy of Sciences, National Research Council, *Compensating for Wetland Losses under the Clean Water Act* (2001) (highlighting significant problems with mitigation efforts to date, including mitigation carried out by the Corps); National Academy of Sciences, National Research Council, *Inland Navigation System Planning: The Upper Mississippi River–Illinois Waterway* (2001) (finding that the Corps used a fundamentally flawed model to assess the lock expansion project and was not properly accounting for the environmental consequences of its plan and recommending that Congress direct the Corps to fully evaluate use of nonstructural measures in its project planning); Department of the Army Inspector General *Investigation of Allegations against the U.S. Army Corps of Engineers Involving Manipulation of Studies Related to the Upper Mississippi River and Illinois Waterway Navigation Systems* (Case No. 00-019) (Nov. 2000) (finding that the Corps deceptively and intentionally manipulated data in an attempt to justify a $1.2 billion expansion of locks on the Upper Mississippi River and that the Corps has an institutional bias for constructing costly, large-scale structural projects); National Academy of Sciences, National Research Council, *New Directions in Water Resources Planning for the U.S. Army Corps of Engineers* (1999) (recommending key changes to the Corps' planning process, including updating the Corps' *Principles and Guidelines*); National Academy of Sciences, National Research Council, *Restoring and Protecting Marine Habitat: The Role of Engineering and Technology* (1994) (finding, among other things, that the Corps should revise its policies and procedures to improve restoration and protection of marine resources).

15. U.S. Water Resources Council, *Economic and Environmental Principles and Guidelines for Water and Land Resources Implementation Studies* (1983) ("the *P&G*"). The *P&G* apply to four principal federal agencies that historically have planned new water projects: the Corps, the Natural Resources Conservation Service, the Bureau of Reclamation, and the Tennessee Valley Authority. Of these agencies, the Corps is the only one that continues to plan new projects in any significant number. The Water Resources Council was established to encourage the conservation, development, and utilization of water and related land resources. Council members included the Secretaries of Agriculture, Army, Commerce, Energy, Housing and Urban Development, Interior, and Transportation and the administrator of the Environmental Protection Agency. National Academy of Sciences, National Research Council, *Analytical Methods and Approaches for Water Resources Planning* 19–20 (2004). In late 1983, the Council's acting chair,

Secretary of the Interior James Watt, deactivated the Council by eliminating its funding and staff, and the Council remains inactive to this day. The Council wrote the *P&G* in 1983 before the Council was deactivated.

16. Benefit-evaluation procedures specific to urban flood damage-reduction projects are described in the Corps' Engineering Regulations. *See, e.g.*, U.S. Army Corps of Engineers, E.R. 1105-2-100 (Apr. 22, 2000), at § E-19.

17. National Wildlife Federation & Taxpayers for Common Sense, *Crossroads: Congress, the Corps of Engineers and the Future of America's Water Resources* 15 (March 2004) (*"Crossroads"*).

18. Federal taxpayers pay a significant percentage of the total cost of Corps projects. Federal taxpayers pay 65 percent of the cost of flood damage-reduction projects. Local sponsors pay the remaining 35 percent of flood-control costs— often through state and/or local taxes. 33 U.S.C § 2213 Fifty percent of the cost of navigation projects are paid for by federal taxpayers with the remaining 50 percent paid for by the Inland Waterways Trust Fund. Federal taxpayers pay 100 percent of the costs of operating and maintaining the inland waterway system. 33 U.S.C. § 2212.

19. Efforts to restore environmental quality, including fish and wildlife enhancement, "shall be deemed to be at least equal to the costs of such measures." 33 U.S.C. § 2284. As a result, a formal benefit-cost analysis is not required for restoration projects.

20. 33 U.S.C. §§2282, 2283(d) (2005). In most cases, the Corps cannot conduct a project study unless that study has been specifically authorized by an act of Congress. The Corps then needs a second authorization from Congress before it can begin constructing any project that it recommends through the study process. Exceptions are projects that can proceed under one of the Corps' continuing-authorities programs. These programs typically are limited to projects that cost $5 million or less and that focus on addressing specific situations. *E.g.*, 33 U.S.C. § 2309a (a continuing-authorities program for small-scale modifications to existing Corps projects for the purpose of improving the environment); 33 U.S.C § 2330 (a continuing-authorities program for small-scale aquatic ecosystem-restoration projects).

21. U.S. Government Accountability Office, *Corps of Engineers, Observations on Planning and Project Management Processes for the Civil Works Program*, GAO-06-529T 12, (Mar. 2006).

22. National Research Council, *New Directions in Water Resources Planning for the U.S. Army Corps of Engineers* 4, 21, 61–63 (1999); National Research Council, *Inland Navigation System Planning: The Upper Mississippi River–Illinois Waterway* 25–28, 53–54 (2001); U.S. Army Inspector General, *Report of Investigation*, Case 00-019 (2000), at 7–8.

23. Michael Grunwald, *Engineers of Power: An Agency of Unchecked Clout*, Washington Post, Sept. 10, 2000, at A1 (*quoting* Bill Hartwig, then regional director, Region 3, U.S. Fish and Wildlife Service).

24. *Principles and Guidelines, supra* note 15, at v, ¶ 6.

25. National Research Council, *New Directions in Water Resources Planning for the U.S. Army Corps of Engineers* 4, 61–63 (1999).

26. *Crossroads, supra* note 17, at 16.

27. *Id.*

28. Claiming project benefits for draining wetlands is at odds with the following five policies:

(1) The implementing regulations of the National Environmental Policy Act and the section 404(b)(1) guidelines, both of which require mitigation for wetland impacts that cannot be avoided. 40 C.F.R. §§ 1508.20, 230.10(d) (2005).

(2) Executive Order 11990, which since 1977 has directed every federal agency to provide leadership and take action to minimize the destruction, loss, or degradation of wetlands and to preserve and enhance the natural and beneficial values in carrying out agency responsibilities. Indeed, this Executive Order specifically compels the Corps to avoid draining, dredging, and filling wetlands. Protection of Wetlands Executive Order (Executive Order 11990), *reprinted in* 42 U.S.C. § 4321 (1977). Executive Order 11990 also provides that each federal agency "shall avoid undertaking or providing assistance for new construction located in wetlands unless the head of the agency finds (1) that there is no practicable alternative to such construction, and (2) that the proposed action includes all practicable measures to minimize harm to wetlands which may result from such use." *Id.* at § 2(a). The term "new construction" is defined to include "draining, dredging, channelizing, filling, diking, impounding and related and any structures or facilities begun or authorized after the effective date" of the Executive Order. *Id.* at § 7(b). The courts have held that Executive Order 11990 is judicially enforceable and should be given the full force and effect of law. *City of Carmel by-the-Sea v. United States Department of Transportation*, 123 F.3d 1142, 1166 (9th Cir. 1997); *City of Waltham v. United States Postal Service*, 786 F. Supp. 105, 131 (D. Mass. 1992). The courts also have found that this Executive Order imposes duties on federal agencies beyond those of NEPA. It requires a specific finding that no practicable alternative to the proposed action exists. *City of Carmel*, 123 F.3d at 1167.

(3) National farm policy and the federal government's significant efforts to take excess and environmentally sensitive croplands out of production and to remove incentives for draining wetlands to enhance crop production. The Food Security Act of 1985 and the Erodible Land and Wetland Conservation Program, 16 U.S.C. §§ 3801 et seq. (1985), encourage the removal of fragile lands from production and provide various opportunities for wetland habitat protection and restoration. A special conservation provision in this act, known as "Swampbuster," removes incentives for draining wetlands by eliminating most agricultural subsidies to farmers who drain wetlands to enhance crop production or who produce commodities on wetlands converted after 1985.

(4) For floodplain wetlands, this practice is at odds with Executive Order 11988, which since 1977 has directed all federal agencies to take action to "restore and preserve the natural and beneficial values served by floodplains" in carrying out their water-resources activities. This Executive Order was passed to help reduce flood damages by protecting the natural values of floodplains and reducing unwise land-use practices in the nation's floodplains.

(5) The long-standing bipartisan national policy of no net loss of wetlands, which was established under the first Bush administration and which was codified as to the Corps in the Water Resources Development Act of 1990. 33 U.S.C. § 2317(a)(1) (1990).

29. National Research Council, *New Directions in Water Resources Planning for the U.S. Army Corps of Engineers* 4–5 (1999); National Academy of Sciences, National Research Council, *Analytical Methods and Approaches for Water Resources Planning* 5 (2004).

30. *See, e.g.,* the studies cited *supra* in note 14. Prior to July 7, 2004, the Government Accountability Office was known as the General Accounting Office, and a number of the GAO reports discussed in this essay were completed prior to that name change. *Available at* http://www.gao.gov/about/namechange.html (last visited June 1, 2005).

31. U.S. Government Accountability Office, *Corps of Engineers, Observations on Planning and Project Management Processes for the Civil Works Program,* GAO-06-529T, at 5 (Mar. 2006).

32. *Id.*

33. U.S. Government General Accounting Office, *Delaware River Deepening Project: Comprehensive Reanalysis Needed,* GAO-02-604, at 4 (June 2002) ("GAO Delaware River Review").

34. *Crossroads, supra* note 17, at 57.

35. GAO Delaware River Review, *supra* note 33, at 5.

36. *Id.*

37. *Id.* at 2.

38. *Crossroads, supra* note 17, at 57 and nn. 225 and 226.

39. *Id.*

40. Michael Grunwald, *A Race to the Bottom,* Washington Post, Sept. 12, 2000, at A1.

41. *Crossroads, supra* note 17, at 79.

42. Department of the Army Inspector General, *Report of Investigation,* Case No. 00-019, Investigation of Allegations against the U.S. Army Corps of Engineers Involving Manipulation of Studies Related to the Upper Mississippi River and Illinois Waterway Navigation Systems, at 1 (Nov. 13, 2000) ("The evidence also indicated that the economic analysis prepared for the draft report was manipulated. The District Engineer (DE) directed a specific value for a key parameter when he knew it was mathematically flawed, not empirically based, and contrary to the recommendations of Corps economists. Evidence also revealed that the former Director for Civil Works (DCW) and the Mississippi Valley Division (MVD) Commander created a climate that led to the manipulation of the benefits-cost analysis").

43. *Upper Mississippi River–Illinois Waterway Feasibility Study: Second Report, supra* note 14, at 7–10 (finding that flaws in the Corps' models used to predict

demand for barge transportation preclude a demonstration that expanding the locks is economically justified and that the Corps has not given sufficient consideration to inexpensive, nonstructural navigation improvements that could ease congestion at the existing locks); *Upper Mississippi River–Illinois Waterway Feasibility Study* (First Report), *supra* note 14, at 3 (finding that the Corps' models are fundamentally flawed, traffic projections are significantly overstated, and the Corps cannot assess the need for the project until small-scale measures to reduce congestion are implemented).

44. *Report of the Chief of Engineers of the U.S. Army Corps of Engineers on the Upper Mississippi River–Illinois Waterway System Navigation and Ecosystem Improvements* 2 (Dec. 15, 2004).

45. Editorial, *Wet Elephant*, Washington Post, Jan. 5, 1987, at A16.

46. 33 U.S.C. § 2283(d) (2005).

47. 33 U.S.C. § 2317(a)(1) (2005).

48. U.S. General Accounting Office, *U.S. Army Corps of Engineers: Scientific Panel's Assessment of Fish and Wildlife Mitigation Guidance*, GAO-02-574, at 4 (May 2002). The Corps provided the GAO with mitigation planning information for 150 projects that the Corps says were authorized between the Water Resources Development Act of 1986 and September 30, 2001, and that received construction appropriations. Only forty seven of those projects (or just 31 percent) included mitigation plans. *Id.*

49. U.S. Army Corps of Engineers, E.R. 1105-2-100, at E-89 (Apr. 22, 2000).

50. The Environmental Protection Agency gave the Corps' environmental impact statement for this project a rating of EU2. The criteria for that rating, which includes the quote referenced in the text, are *available at* http://www.epa.gov/compliance/nepa/comments/ratings.html (last visited May 28, 2005).

51. The Environmental Protection Agency gave the Corps' environmental impact statements for each of these projects a rating of EO2. The criteria for that rating, which includes the quote referenced in the text, are *available at* http://www.epa.gov/compliance/nepa/comments/ratings.html (last visited May 28, 2005).

52. Information supporting the GAO's May 2002 study entitled *U.S. Army Corps of Engineers: Scientific Panel's Assessment of Fish and Wildlife Mitigation Guidance*, GAO-02-574 (May 2002). The list of projects was provided to American Rivers by the U.S. Army Corps of Engineers on request.

53. To ensure timely implementation of mitigation, the Corps is required to implement mitigation before constructing a civil-works project with just one exception. Mitigation that requires physical construction can be carried out concurrently with project construction. 33 U.S.C. § 2283(a) (2005). Despite this requirement, the GAO found that in those 31 percent of cases where the Corps is mitigating the harm from its projects, it is failing to mitigate prior to or concurrently with project construction 80 percent of the time. *Assessment of Fish and Wildlife Mitigation Guidance, supra* note 52, at 4.

54. *Final Project Report and Supplement No. 2 to the Final Environmental Impact Statement, Flood Control, Mississippi River and Tributaries, Yazoo*

Basin, Mississippi, Big Sunflower River Maintenance Project, vol. 1, *Project Report, Supplemental Environmental Impact Statement and Appendices A–C* (July 1996), at app. B, U.S. Fish and Wildlife Coordination Act Report at i. At least 443 acres of bottomland hardwood wetlands and 476 acres of farmed wetlands will be destroyed, and an additional 2,712 acres of wetlands will be severely impacted. *Id.*

55. U.S. Army Corps of Engineers, Vicksburg District, response to Freedom of Information Act Request No. 00-60 submitted by Melissa Samet, Earthjustice (Nov. 7, 2000). The request sought information and data on the Corps' wetlands monitoring program in the Vicksburg District.

56. *E.g.*, U.S. States Geological Survey, News Release, *Without Restoration, Coastal Land Loss to Continue* (May 21, 2003), *available at* http://www.nwrc.usgs.gov/releases/pr03_004.htm (last visited May 22, 2006).

57. *E.g.*, Sullivan, *supra* note 7.

58. Marshall, *supra* note 7.

59. Matthew Brown, *Katrina May Mean MR-GO Has to Go, Channel Made Storm Surge Worse, Critics Say,* New Orleans Times Picayune, Oct. 24, 2005, *available at* http://www.nola.com/news/t-p/frontpage/index.ssf?/base/news-4/1130133302133590.xml (last visited May 23, 2006); Michael Grunwald, *Canal May Have Worsened City's Flooding*, Washington Post, Sept. 14, 2005, at A21; Louisiana State University, Louisiana Coast, and Sea Grant Louisiana, *"Closing" the Mississippi River Gulf Outlet, Environmental and Economic Considerations*, *available at* http://www.ccmrgo.org/documents/closing_the_mrgo.pdf (last visited Sept. 21, 2005); Personal communication with G. Paul Kemp, Director, Natural Systems Modeling Group, Louisiana State University (2005).

60. Brown, *supra* note 59; Grunwald, *supra* note 59; Louisiana State University, *supra* note 59.

61. Jim Barnett, *Instead of Shoring Up Levees, Corps Built More,* The Oregonian, Sept. 18, 2005, *at* http://oregonlive.com (last visited September 19, 2005).

62. Editorial, *After the Flood, Is New Orleans Safe?*, The Register-Guard (Eugene, OR), Mar. 5, 2006, *available at* http://www.registerguard.com/news/2006/03/05/printable/ed.edit.return.0305.hr19yUF4.phtml?section=opinion (last visited Mar. 13, 2006).

63. Helen Lambourne, *New Orleans "Risks Extinction,"* BBC News, Feb. 3, 2006, available at http://news.bbc.co.uk/go/pr/fr/-/2/hi/science/nature/4673586.stm (last visited Mar. 13, 2005).

64. *E.g.*, Testimony from the November 2, 2005, Senate Homeland Security and Government Affairs Committee hearing, *available at* http://hsgac.senate.gov/index.cfm?Fuseaction=Hearings.Detail&HearingID=290 (last visited Nov. 11, 2005); Bob Marshall, *Floodwall Failure Was Foreseen, Team Says*, New Orleans Times Picayune, Mar. 14, 2006, *available at* http://www.nola.com/printer/printer.ssf?/base/news-5/1142320591212390.xml (last visited Nov. 15, 2006); Kris Axtman, *Search for Weak Link in Big Easy's Levees*, Christian Science

Monitor, Dec. 30, 2005, *available at* http://www.csmonitor.com/2005/1230/
p03s03-sten.html (last visited Nov. 15, 2006); Lisa Myers, *New Orleans Levee
Reported Weak in 1990s*, MSNBC, Sept. 30, 2005, *available at* http://www
.msnbc.msn.com/id/9532037/ (last visited Oct. 5, 2005); Ken Kaye, *Katrina May
Have Been a Category 3 Hurricane, Not 4, When It Struck New Orleans*, South
Florida Sun Sentinel, Oct. 4, 2005, *at* http://www.sun-sentinel.com/news/local/
southflorida/sfl-skatrina04oct04,0,2218266.story (last visited Oct. 4, 2005);
Michael Grunwald & Susan B. Glasser, *Experts Say Faulty Levees Caused Much
of Flooding*, Washington Post, Sept. 21, 2005, at A01; Christopher Drew &
Andrew C. Revkin, *Design Shortcomings Seen in New Orleans Flood Walls*,
New York Times, Sept. 21, 2005, *available at* http://www.nytimes.com/2005/
09/21/national/nationalspecial/21walls.html?pagewanted=2 (last visited Nov. 15,
2006).

65. Bob Marshall & Mark Schleifstein, *Corps Ignored Crucial Levee Data,
Reports Showed Need for Higher Defenses*, New Orleans Times Picayune, Mar.
8, 2006, available at http://www.nola.com/news/t-p/frontpage/index.ssf?/base/
news-5/1141802754126640.xml (last visited Mar. 13, 2006).

66. Bill Walsh, *Corps Chief Admits to "Design Failure,"* New Orleans Times
Picayune, Apr. 6, 2006, *available at* http://www.nola.com/news/t-p/frontpage/
index.ssf?/base/news-5/1144306231230500.xml (last visited Apr. 26, 2006).

67. *E.g.*, Sheila Grissett, *Katrina Report Blames Human Errors, Poor Levees,
Policies Cited by Investigators*, New Orleans Times Picayune, May 22, 2006,
available at http://www.nola.com/news/t-p/frontpage/index.ssf?/base/news-5/
1148278528120750.xml (last visited May 23, 2006); John Schwartz, *New Study
of Levees Faults Design and Construction*, New York Times, May 22, 2006,
available at http://www.nytimes.com/2006/05/22/us/22corps.html?_r=1&
oref=slogin (last visited May 23, 2006); Letter from the American Society of Civil
Engineers External Review Panel to LTG Carl A. Strock, Chief of Engineers,
regarding the External Review Panel Progress: Report Number 1 (Feb. 20, 2006).

68. S. 2288, Water Resources Planning and Modernization Act of 2006, 109th
Cong., 2d Sess. (Feb. 15, 2006); S. 753, Corps of Engineers Modernization and
Improvement Act of 2005, 109th Cong., 1st Sess., (Apr. 11, 2005); S. 2188, Corps
of Engineers Modernization and Improvement Act of 2004, 108th Cong. (Mar. 10,
2004); S. 3036, Corps of Engineers River Stewardship Independent Investigation
and Review Act, 107th Cong. (Oct. 3, 2002); S. 2963, Corps of Engineers Reform
Act of 2002, 107th Cong. (Sept. 18, 2002); S. 1987, Corps of Engineers Modernization and Improvement Act of 2002, 107th Cong. (March 5, 2002); S. 646,
Corps of Engineers Reform Act of 2001, 107th Cong. (March 29, 2001); H.R.
2566, Army Corps of Engineers Reform Act of 2003, 108th Cong. (June 23, 2003);
H.R. 2353, Army Corps of Engineers Reform and Community Relations Improvement Act of 2001, 107th Cong. (June 27, 2001); H.R. 1310, Corps of Engineers Reform Act of 2001, 107th Cong. (March 29, 2001).

69. *Civil Works Program Statistics Fact Sheet, supra* note 2; *Crossroads, supra*
note 17, at 29 and n. 105.

70. Water Resources Development Act of 2000, Pub. L. No. 106–541 (Dec. 11,
2000).

71. U.S. General Accounting Office, *Improved Analysis of Costs and Benefits Needed for Sacramento Flood Protection Project,* GAO-04-30, at 1 (Oct. 2003) (*"GAO Sacramento Project Review"*).

72. *Id.*

73. *Id.* at 1–2, 7.

74. *Id.*

75. *Id.* at 1–2.

76. *Id.* at 4.

77. *Id.* Cost overruns on Corps projects are not unusual. For example, the Corps' estimate for just four of the many projects in its Everglades ecosystem restoration program increased by almost $1 billion in just five years. Gary Hardesty, U.S. Army Corps of Engineers, Five-Year Report to Congress, HQUSACE Guidance, March 7, 2005 (this internal memorandum was leaked to the press, a copy is on file with the author and is *available at* http://www .peer.org/docs/ace/2005_21_3_everglades_5-year_report.pdf (last visited June 1, 2005).

78. *GAO Sacramento Project Review, supra* note 71, at 5.

79. *Id.* at 21.

80. *Id.* at 19–20.

81. *Id.* at 4.

82. *Dallas Floodway Extension Project Description, available at* http://www .swf.usace.army.mil/pubdata/pao/dfe/index.asp (last visited May 27, 2005).

83. *Record of Decision, Environmental Impact Statement, Dallas Floodway Extension, Texas* 1 (Dec. 1, 1999), *available at* http://www.swf.usace.army.mil/ pubdata/pao/dfe/DFERecordOfDecision.asp (last visited May 28, 2005).

84. *Dallas Floodway Extension Project Description, available at* http://www. swf.usace.army.mil/pubdata/pao/dfe/index.asp (last visited May 27, 2005); *Record of Decision, supra* note 83; Trinity River Corridor Project, *Fact Sheet: The Corps' Dallas Floodway Extension Project* (May 10, 2005), *available at* http://www.trinityrivercorridor.org/pdf/DFEFactsheet.pdf (last visited May 27, 2005); *Crossroads, supra* note 17, at 60.

85. *Crossroads, supra* note 17, at 60.

86. U.S. Army Corps of Engineers, Ft. Worth District, *Final Supplement 1 to Environmental Impact Statement for the Dallas Floodway Extension, Trinity River Basin, Texas* 2–4 (April 2003), *available at* http://www.swf.usace.army.mil/ pubdata/pao/dfe/pdfs/FINAL%20SEIS%20Main%20Text.pdf (last visited May 27, 2005); *Record of Decision, supra* note 83, at 2.

87. *Final Supplement 1, supra* note 86, at ix. The U.S. District Court for the Northern District of Texas enjoined construction of the Dallas Floodway Extension in April 2002 until the Corps completed an adequate cumulative impacts assessment of the project. Dallas Floodway Extension Project Description, *available at* http://www.swf.usace.army.mil/pubdata/pao/dfe/index.asp (last visited May 27,

2005). That injunction was lifted on May 5, 2004. Trinity River Corridor Project, *Fact Sheet, The Corps' Dallas Floodway Extension Project* (May 10, 2005), *available at* http://www.trinityrivercorridor.org/pdf/DFEFactsheet.pdf (last visited May 27, 2005); Dallas Floodway Extension Project Description, *available at* http://www.swf.usace.army.mil/pubdata/pao/dfe/index.asp (last visited May 27, 2005).

88. Trinity River Corridor Project, *supra* note 84.

89. U.S. Army Corps of Engineers, *Initial Assessment, Dallas Floodway Extension Reevaluation of the Cadillac Heights Floodplain Evacuation Measure* 9 (June 14, 2001), *available at* http://www.swf.usace.army.mil/pubdata/pao/dfe/pdfs/ASA_DFE_IA_Report_Final1_Encl.pdf (last visited May 27, 2005).

90. Letter from Mitchell E. Daniels, Jr., Director, Office of Management and Budget, to Thomas E. White, Secretary of the Army (Oct. 3, 2001), at 1, *available at* http://www.swf.usace.army.mil/pubdata/pao/dfe/pdfs/DFE_Daniels_Ltr_to_White.pdf (last visited May 27, 2005).

91. *Id.* at 1–2; *Crossroads, supra* note 17, at 60.

92. *Crossroads, supra* note 17, at 60 (*citing* Fort Worth District, U.S. Army Corps of Engineers, Dallas Floodway Extension Project, *Information Paper: Dallas Floodway System Phasing* 5, tab. 3 (Aug. 3, 2001) (estimating that raising the existing East Dallas Levee would cost approximately $21.7 million at 1998 price levels, compared with a $110 million estimate for the DFE project), *available at* http://www.swf.usace.army.mil/pao/dfe/pdfs/ASA_DFE_OMB_Analysis_Encl.pdf (last visited Sept. 30, 2003); and *also citing* Fort Worth District, U.S. Army Corps of Engineers, *Initial Assessment: Dallas Floodway Extension Reevaluation of the Cadillac Heights Floodplain Evacuation Measure* 5, tab. 2 (June 14, 2001), *available at* http://www.swf.usace.army.mil/pao/dfe/pdfs/ASA_DFE_IA_Report_Final1_Encl.pdf (last visited Sept. 30, 2003).

93. American Rivers, *America's Most Endangered Rivers of 2003*, at 32, *available at* http://www.americanrivers.org/site/PageServer?pagename=AMR_content_01bd (last visited May 28, 2005).

94. Uolaku Echebiri, *A Summary of the Napa Flood Control Project*, Institute for Crisis, Disaster, and Risk Management, Crisis and Emergency Management Newsletter, Jan. 2004, *available at* http://www.seas.gwu.edu/~emse232/january2004mitigation1.html (last visited May 28, 2005).

95. An earlier project had been authorized in the Flood Control Act of 1944, but Congress never appropriated funds to carry it out. The City of Napa instead constructed a portion of that project, but it did not stop the city's flooding problems. Napa Flood and Water Conservation District, *Creating Flood Protection, The History of Floods and the Community Coalition*, *available at* http://www.napaflooddistrict.org/FloodDetail.asp?LID=550 (last visited May 28, 2005).

96. *Id.*

97. *Id.*

98. *Id.*

99. *Id.*

100. *Id.*

101. *Id.*; Napa River Fisheries Monitoring Program, *available at* http://www
.napariverfishmonitoring.org/history/history.html (last visited May 28, 2005).

102. B. Otto, K. McCormick, & M. Leccese, *Ecological Riverfront Design:
Restoring Rivers, Connecting Communities* (2004), also *available at* http://www
.americanrivers.org/site/PageServer?pagename=AMR_content_2bb4 (last visited
May 26, 2005).

103. *Id.* at 47.

104. M. A. Palmer, E. S. Bernhardt, J. D. Allan, et al., *Standards for Ecologically
Successful River Restoration*, 42 J. Applied Ecology, 208 (April 2005).

105. Congress appropriated $1.8 billion for Corps construction in fiscal year
2003, $1.7 billion for Corps construction in fiscal year 2004, $1.8 billion for
Corps construction in fiscal year 2005, and $2.4 billion for Corps construction in
fiscal year 2006.

106. For example, the removal of two small dams and the realignment of the
channel of Wyomissing Creek at the Reading Museum property in eastern
Pennsylvania was carried out by a private contractor at total cost of $300,000,
with dam removals accounting for about $60,000 of that total. The Philadelphia
District of the Corps provided the Reading Museum with a preliminary estimate
of $1.3 million for the same work. Design work was not included in the Corps'
estimate. Personal communication with Sara Nicholas, American Rivers. The
$300,000 total cost of the privately constructed project is far below the 35 per-
cent cost share of $445,000 that would have been required for a $1.3 million
restoration project constructed by the Corps.

8

Bankside Citizens

Mike Houck

In Defense of Trashed Rivers

When we initiated the Urban Naturalist Program at the Audubon Society of Portland in Portland, Oregon, in the summer of 1982, I was stunned to discover that local planners considered the notion of city planning that protects or restores fish and wildlife habitat and provides "nature in the city" as oxymoronic. In fact, efforts to protect and restore urban natural resources were viewed as antithetical to "good planning" and an affront to their urban design aesthetic. I was told good-intentioned but ill-informed environmentalist do-gooders' efforts to protect natural areas in our region was quixotic and contrary to Oregon's land-use planning program, which was designed to protect nature "out there" in the rural hinterlands, not in the city. There was, in short, no room for nature in cities, which are meant to be high density, human dominated, and the exclusive domain of the built landscape.

In the intervening twenty-three years, planners and even many environmental organizations have reconsidered their attitudes toward retaining nature in the city. Nevertheless, there are still residual pockets of antiurban nature sentiment within the New Urbanist and smart-growth camps. Smart-growth policies such as Urban Growth Boundaries have, to their credit, been a boon to curbing urban sprawl. In the Portland metropolitan region, for example, population expanded by 31 percent between 1990 and 2000, and land consumption increased by only 3 percent. That is a phenomenal containment of urbanization onto prime farmland and forest land. It also flies in the face of the opposite trend in the rest of the United States, where even areas like the Syracuse, New York, region, which has lost 50 percent of its population, continue to sprawl.

However, Portland's lack of commitment to protecting natural resources within its Urban Growth Boundary has resulted in an inequitable distribution of parkland, loss of natural resources, ditching, culverting and burying of urban streams, degraded water quality, and disappearance of fish and wildlife habitat. Portland has over 213 miles of streams that have been deemed "water-quality limited" (or in common language, polluted) by the Oregon Department of Environmental Quality. And more than 400 miles of streams have been put underground in culverts to accommodate regional growth.

The negative impacts to urban streams and other natural resources was documented in the *The Oregon State of the Environment Report, 2000*,[1] which concluded: "The annual rate of conversion of forest and farmlands to residential and urban uses has declined dramatically since comprehensive planning land use planning was implemented during the 1980s. However, these laws were not written to address ecological issues, such as clean water or ecosystem function within urban growth boundaries. In order to meet the economic and social needs of humans, native vegetation and habitats may be destroyed and converted to buildings and paved surfaces."

The Defenders of Wildlife's 2001 report *No Place for Nature: The Limits of Oregon's Land Use Program in Protecting Fish and Wildlife Habitat in the Willamette Valley* also reveals serious shortcomings with regard to urban natural resource protection: "The land use program . . . tends to focus on one goal, one resource at a time. Furthermore, the land use program is implemented by a multitude of local governments. At present, the planning program carries no requirement that these entities coordinate their approaches to, for example, riparian corridors that may extend across several jurisdictions."[2]

For a variety of reasons, by the late 1970s and early 1980s the urban planners and engineers charged with building our cities seemed to have either forgotten or willfully abandoned the tenets of the City Beautiful movement and Ian McHarg's "design with nature" precepts. By the 1980s, few city planners were heeding Anne Whiston Sprin's admonition in her 1984 book *The Granite Garden: Urban Nature and Human Design* to think in radically different ways about how cities are built. In *The Granite Garden*, Spirn states, "The belief that the city is an entity apart from nature and even antithetical to it has dominated the way in which the city is perceived and continues to affect how it is built. The city must be recognized as part of nature and designed accordingly."[3]

Forgotten or tossed aside as romantic and naïve were the turn-of-the-century urban greenprints from park and greenspace visionaries such as Frederick Law Olmsted,[4] H.W.S. Cleveland,[5] and Charles Eliot[6]—all of whom understood and advocated on behalf of creating humane cities by integrating streams, wetlands, and the natural landscape within the urban fabric.

In Portland, Oregon, John Charles Olmsted presaged the concepts of multiobjective river-corridor management and modern watershed restoration when he called for the creation of a comprehensive, interconnected park system in his 1903 master plan for the city: "Marked economy in municipal development may be effected by laying out parkways and parks, while land is cheap, so as to embrace streams that carry at times more water than can be taken care of by drain pipes of ordinary size. Thus, brooks or little rivers that would otherwise become nuisances that would some day have to be put in large underground conduits at enormous expense, may be made the occasion for delightful local pleasure grounds or attractive parkways. Such improvements add greatly to the value of adjoining properties, which would otherwise been depreciated by the erection on the low lands of cheap dwellings or by ugly factories, stables and other commercial establishments."[7] Would that we had heeded Olmsted's sage advice, given that a large part of solving our $1.4 billion price tag for Portland's combined sewer overflow (CSO) program to remove raw sewage from the Willamette River would have been far less costly had we not buried over 300 miles of streams[8] in the intervening 100 years.

From the time the Oregon planning program[9] was adopted in the early 1970s to address containment of urban sprawl and protection of farm and forest land until very recently, few efforts have been made to integrate "nature in the city" into urban planning. By the 1980s, there were a few notable exceptions, including David Goode and his urban ecology team in the London Ecology Unit[10] and the National Institute for Urban Wildlife's efforts to encourage planning for urban wildlife through national conferences and publications.[11]

Even in relatively progressive Portland, Oregon, scant attention was paid to urban stream and natural-resource protection, and urban waterways continued to be degraded nationally.

A Grassroots Movement for Trashed Rivers

In that context locally and nationally, a national grassroots movement to protect and restore urban waterways emerged in the late 1980s and early 1990s. That movement, decidedly grassroots in nature, had many loci and arose spontaneously across the country. One of the catalysts was a Country in the City[12] symposium in Portland, Oregon, at which A. L. Riley (cofounder of California's Urban Creeks Council in Berkeley), Robbin Sotir (of Robin Sotir & Associates in Marietta, Georgia), and Joan Florsheim (of Philip Williams Associates in San Francisco) all made presentations. Each managed to ruffle feathers, especially among the male-dominated waterway engineering professionals in attendance. Riley, Sotir, and Florsheim had all been invited to speak at the suggestion of Jon Kusler, who was then director of the National Association of State Wetland Managers and was actively involved with the National Association of State Floodplain Managers. All three women brought a strong penchant—based on years of practical, on-the-ground experience—for nonstructural, "soft-engineered" alternatives for restoring urban streams and rivers. Those of us who had long preached nonstructural approaches to urban stream and watershed management had finally found someone who could, with engineering credentials, articulate a vision for protecting and restoring urban waterways.

Most great ideas, in my experience, originate over good food and drink. A case in point was the national urban-streams movement. Ann Riley's proposal for such a movement emerged over green curried chicken and Singha beer in west Berkeley shortly after the 1990 Country in the City symposium. It was clear from her provocative talk that she had a strong command of national floodplain management and waterway policies as well as the technical expertise to make the case for alternative methods of floodplain and urban-waterway management. Riley drew out on the paper placemat at our table what would become a road map, a route to creating a grassroots coalition of urban-stream restorationists.

The conversation grew more animated as Riley's arrows and annotations spread across the disposable paper mat. What she drew was nothing short of a new vision for a national policy for urban-stream and river-restoration and management. Her scribblings established connections among myriad federal and state programs, including the Natural

Resources Conservation Service's emerging urban-stream initiative and the National Park Service's Rivers and Trails Conservation Assistance Program,[13] which had a strong "multiobjective"[14] river-management initiative for both rural and urban waterways. The predilection for seeking multiple benefits among these federal agencies fit nicely into Riley's vision for restoring the multiple values associated with streams and rivers in the most urbanized and (more often than not) most economically depressed communities. Combined with the Soil Conservation Service and National Park Service programs Riley was seeking ways to harness the professional and political savvy among the engineers and river planners in the Associations of State Wetland Managers and sister group the Association of State Floodplain Managers.

Once Riley had connected the dots, the time was ripe to take her map on the road. Her long-term experience in urban-stream restoration as well as her national stature (based on friendships with Jon Kusler at the Wetland Managers, Chris Brown at the National Parks Service, and her Berkeley mentor Luna Leopold; on a high-profile role on National Academy of Sciences national advisory panels; and on professional relationships with researchers at the U.S. Army Corps of Engineers' Waterways Research Center) made her the logical leader around whom to coalesce a national grassroots urban stream-restoration movement.

Assembling CRUW's Crew: Berkeley, 1993

Shortly after our Thai restaurant map session, Riley, in her capacity as director of the Golden State Wildlife Federation, invited a number of agency staff and representatives from nonprofits and community groups to the recently "daylighted" Strawberry Creek in west Berkeley.[15] Among the February 25–26, 1993, invitees and participants were California Congressman George Miller and Chris Brown of the National Park Service Rivers and Trails Conservation Assistance Program in Washington, D.C. Also present were Karen Firehock, director of the Izaak Waltons League's Save Our Streams program; Esther Lev and myself, then representing the Urban Stream Council in Portland, Oregon; Nancy Stone, National Park Service Rivers and Trails Program; Beth Stone, Urban Creeks Council of California; Perle Reed, Soil Conservation Service; Mike Leedie, North Richmond Neighborhood; Walter Hood, Urban Creeks Council; Elizabeth

Hertel, Indiana Dunes Environmental Education; Larry Fishbain, Philip Williams Associates; Jon Kusler, Association of State Wetland Managers; Jack Byrne, River Watch Network; Lillie Mae Jones, a Richmond, California, community activist; Martin Schlageter, Friends of the Los Angeles River; Bob Sneikus, Soil Conservation Service; Bob Doppelt, Pacific Rivers Council, Oregon; Pete Lavigne, River Network; Beth Stone, Urban Creeks Council, Berkeley, California; Nancy Stone, National Park Service Rivers and Trails program; Larry Fishbain, Philip Williams Associates; Eugenia Laychak, Coastal Resources Center, California; and Deborah Alex Sanders, Minority Environmental Organization.

This geographically, organizationally, and ethnically diverse group— by far the most racially and ethnically diverse group I had seen gathered around an environmental issue—reflected both the powerful attraction that urban waterways hold across socioeconomic and demographic lines and the enormous respect Riley commanded among her peers as well as government bureaucrats. No movement rises from a single leader, but each invariably has a charismatic leader at its heart. Riley was clearly such a leader in the emerging national urban-streams movement.

The case was clearly made at this founding meeting that this was a propitious time to create a new national, grassroots-based organization that would focus on the restoration and management of urban streams. None of the national environmental organizations were working in any serious way on urban streams. There was a strong sentiment among those gathered at Strawberry Creek that the urban-streams movement would focus on urban waterways in general regardless of their location but that this new organization would pay particular attention to streams and rivers in low-income, economically depressed areas. This nascent urban-stream movement set its roots deeply and resolutely within an environmental-justice matrix. The founding members were passionately insistent that their work be as much about restoring both communities and the streams that ran through them. They also insisted that, contrary to the wishes of some national environmental organizations, the Department of Agriculture's Natural Resources Conservation Service (not the Department of Interior, where the national environmental groups' contacts mostly lay) was the logical home for our efforts to write new national environmental legislation to provide the resources that the grassroots needed to do their work. This schism resulted in tensions between the grassroots urban-streams movement and some mainstream environmental groups.

As with every founding of a nongovernmental organization, we spent a great deal of time deciding what we would call ourselves. Federation of Urban Creek Councils came up early in the process but was quickly discarded because there was some queasiness, especially among our potential federal partners, with the presumed acronym. We eventually settled on the Coalition to Restore Urban Waters (CRUW). The next task, again following the typical course of new movements, was to develop a mission and objectives. On February 26, 1993, we adopted CRUW's Statement of Purpose, which read as follows:

The Coalition to Restore Urban Waters (CRUW) is a national network of diverse grassroots groups which protect and restore urban watersheds, waterways, and wetlands. The Coalition includes all peoples and groups, including ethnically diverse and disenfranchised interests, local and state conservation corps and service corps, educational institutions, nonprofit creeks councils, conservation groups, and citizens committed to restoration of urban waters.

The Coalition works with local communities to address the unique values, opportunities, and issues of urban waterways. Urban waterways are an important link between the environment, the economy, recreation, and neighborhood identity in the community. While the Coalition focuses on urban ecosystems, it recognizes the connection among urban environments and rural, suburban, and wildlands watersheds.

This Coalition was established to provide its partners with

· Networking and information sharing;
· Technical assistance and successful restoration models;
· Promotion of economic opportunities through restoration of urban waters;
· Assistance with funding opportunities;
· A forum for collaboration among traditionally defined environmental groups and disenfranchised urban populations;
· Opportunities for environmental education, curricula, community awareness, and environmental stewardship; and
· A forum for partnerships between grassroots groups and national environmental groups, fisheries groups, local, state, and federal agencies, peace corps, service and conservation corps, and business interests.

Our first project, we decided, would be to host a national urban-streams conference for the grassroots partners that CRUW was created to serve. The purpose of the conference would be to attract urban stream and wetland advocates from around the United States and beyond and to get their read on the need for a national grassroots group like CRUW. The steering committee for planning the conference included A. L. Riley and Dennis O'Connor, representing the Golden State Wildlife Federation; Esther Lev, Urban Streams Council of the Portland-based Wetlands Conservancy; Mike Houck, Audubon Society of Portland; Karen Firehock, Isaak Walton League's Save Our Streams program; and Carole Schemmerling, a cofounder of the Urban Creeks Council of California.

In keeping with the somewhat irreverent nature of CRUW, the steering committee chose as the conference title Friends of Trashed Rivers, which seemed fitting given the committee's determination to focus its efforts on the restoration of the most degraded waterways in the United States—the Anacostia, the Los Angeles River, the Chicago River, and the Columbia Slough in Portland. We felt that the conference title would attract activists but not be off-putting to potential funders and federal agencies that understood our objectives.

Friends of Trashed Rivers I: San Francisco, 1993

CRUW's grassroots reputation was assured by the 1993 conference. Restoring Urban Waters, Friends of Trashed Rivers: A Conference of the Coalition to Restore Urban Waters was held at the Fort Mason Center in San Francisco on September 17–19, 1993, and drew 300 hard-core urban-stream and wetland-restoration advocates from across the United States and Canada. The conference was sponsored by a number of urban waterway and wetland groups from across the country: the Isaak Walton League's Save Our Streams program; California's Urban Creeks Council; the Urban Streams Council, Portland, Oregon; REI, Inc.; American Rivers; the River Network; the Friends of the Los Angeles River; the Friends of the Chicago River; the Friends of the White River; the Minority Environmental Council; the National Association of Service and Conservation Corps; the Adopt-a-Stream Foundation, Seattle, Washington; the Association of State Floodplain Managers; and the Association of State Wetlands Managers.

One of the hallmarks of the Friends of Trashed Rivers I conference was that it was supported financially by federal agencies (including the U.S. Environmental Protection Agency Region IX; EPA's Office of Wetlands, Oceans, and Watersheds in Washington, D.C.; the Natural Resources Conservation Service; the Bureau of Reclamation; the Tennessee Valley Authority; the National Park Service's Rivers and Trails Conservation

Figure 8.1
Participants at the Friends of Trashed Rivers I conference, San Francisco, 1993.
Photograph provided by A. L. Riley.

Assistance Program; and the U.S. Army Corps of Engineers) but no federal or state representatives were invited or allowed to make presentations. This decision is key to understanding the philosophical goals of CRUW.

While we wanted to partner with the federal agencies and mainstream national environmental groups, we also wanted to honor and ensure that the true "bottom feeders" of the urban conservation movement—urban-stream advocates—were placed in the limelight. We were determined that the focus, for once, would be on the small, mostly all-volunteer community-based nonprofits, not on the federal agencies and the national environmental community. The steering committee appreciated the federal bureaucracies' fiscal support and the support from mainstream national environmental organizations but emphasized that the conference was to be a conference of the grassroots, for the grassroots, and by the grassroots. There were a few bruised egos among some agencies and large environmental groups, but most participants understood our objective of giving the stage to the grassroots for a change.

Friends of Trashed Rivers I had a few quirky moments. For example, to ensure that everyone had an opportunity to share a story with conference attendees, a trashcan lid was beaten with a spoon to signal "Your time's up." The image of A. L. Riley, with her large metal spoon poised for a clang, strongly encouraged presenters to wrap up their remarks in a timely fashion. In addition to the formal presentations and breakout sessions, we provided everyone representing grassroots organizations with a ten-minute opportunity to "witness" before the entire assembly.

Those who attended Friends of Trashed Rivers I will remember vividly that as their time expired A. L. Riley would raise a spoon menacingly above the ever more battered trashcan lid. It took only one or two of Riley's deafening "gongs" to persuade speakers to adhere to their allotted time in the limelight.

CRUW: A National Grassroots Coalition

It's one thing to bring 300 urban-stream advocates together for a raucous revival meeting in San Francisco. The 1993 Friends of Trashed Rivers I confence would have been a significant contribution to urban river and stream restoration had CRUW done nothing more than hold that one event. Fortunately, those who planned the conference and several attendees put a more formal, organized face on the Coalition to Restore Urban Waters.

Through a series of conference calls and face-to-face meetings, CRUW's steering committee[16] forged a formal structure for the creation of a national coalition. Although after several years it became necessary to incorporate as a separate section 501(c)(3) nonprofit organization, CRUW was a loose-knit coalition of like-minded nonprofit organizations that were dedicated to focusing resources on urban-stream protection and restoration.

National Legislation, Workshops, and Newsletters

One of CRUW's most important contributions was to sponsor more Friends of Trashed Rivers conferences in New Orleans, Chicago, and New York City. The Coalition to Restore Urban Waters also lobbied heavily in Washington, D.C., where member organizations met with then Senate Appropriations Committee chair, Mark O. Hatfield (R-Oregon), and Congresswoman Elizabeth Furse (D-Oregon) to amend the Natural

Resources Conservation Service's (NRCS) Public Law 566 to make it possible to use existing funds within NRCS to restore urban waterways. President Clinton called for $100 million in funding for watershed programs in the fiscal year 1996 federal budget, and CRUW sought to secure some of those funds for urban-stream programs.[17] The Coalition sent representatives from several member organizations and garnered support from large national environmental organizations as well.

Another of CRUW's priorities was the production of technical-assistance materials for its member organizations. In February 1996, CRUW published its *Urban Waterway Restoration Training Manual* for Youth Service and Conservation Corps. The manual, with accompanying slides, was written by A. L. Riley and Moira McDonald of the Waterways Restoration Institute and was published by another Coalition member organization, the National Association of Service and Conservation Corps.

The Izaak Walton League's Save Our Streams program provided the infrastructure that kept the loose-knit coalition together in its formative years. In 1995, Karen Firehock and Julie Vincentz, acting on behalf of the League as CRUW's coordinator, created a Five-Year Long-Range Plan[18] assisted the CRUW Steering Committee[19] and then helped create the articles of incorporation when the CRUW steering committee decided that to operate over the long term it needed to become its own section 501(c)(3) nonprofit organization. In August 1997, Firehock drafted bylaws, and officers and a board of directors were selected from the existing steering committee and elected.[20]

Where Did CRUW Go?

People frequently ask, "Where did CRUW go?" In a word, everywhere. While CRUW as a formal entity gradually dissolved after the Friends of Trashed Rivers conferences, its mission, esprit de corps, and irreverent spirit are being carried on by its founding individuals and organizations and the hundreds of new urban-stream and river-restoration groups that were created as a direct result of CRUW's Friends of Trashed Rivers conferences, workshops, field trainings, national newsletters, and legislative work. CRUW suffered the fate of many grassroots movements—founders' fatigue, personality clashes, changes in philosophy, and mission drift. But more than anything, CRUW served its purpose. CRUW drew national attention to the cause of urban streams, and its founders moved on to implement the vision articulated at CRUW's founding.

Based on conversations I've had with Riley and others close to CRUW, it's clear that the coalition did change the face of federal government programs during the Clinton years in which the Natural Resources Conservation Service (NRCS), EPA, National Park Service, and Bureau of Reclamation all set up new programs at CRUW's behest. These agencies also started to provide small to moderate-sized grants and funding to support urban-stream projects. Coordinated "store-front" federal urban environmental programs were set up in Chicago, Los Angeles, Atlanta, San Francisco, and other cities across the country. NRCS, for example, provided the financial backing for Riley's book, *Restoring Streams in Cities: A Guide for Planners, Policymakers, and Citizens.*[21] These agencies also provided support for California inner-city environmental youth programs, many of which focused on local waterway issues. Congresswoman Elizabeth Furse's urban-streams bill made it to a Senate-House conference committee where it failed to pass by a few votes. Nevertheless, as a result of CRUW's work with Congresswoman Furse, the NRCS, traditionally associated with rural programs, embraced CRUW's urban agenda enthusiastically. Much of its enthusiasm related to CRUW's explicit nexus between urban-stream restoration and environmental justice, a concept that resonated with NRCS's leadership and its own commitment to diversity.

Riley has observed that the grassroots revival of urban-stream restoration occurred at a time when the national environmental movement, as chronicled in Mark Dowie's book *Losing Ground*, was losing its engaged grassroots membership—some argued because of its overemphasis on direct-mail solicitation of memberships. This phenomenon would play out later in our own experiences with the National Audubon Society, which in my opinion has all but abandoned its commitment to grassroots activism at the local chapter level. In Riley's case, alienation from the National Wildlife Federation was an incentive to become active in the founding and implementation of CRUW and its mission. It's our shared opinion that both the need for and success of CRUW was for some of us a reflection of a desire for true "bottom-up" grassroots activism that was not universally appreciated in the national environmental movement.

CRUW also represented the cutting edge in the evolution of the environmental movement toward more inclusiveness and toward addressing issues of racism and environmental justice. Urban-based conservation corps became one of CRUW's most important core activities.

Another reason that CRUW members returned to grassroots activism was the political shift that the second Bush administration brought to federal environmental policies. The same is true at the state level in some states. When an overtly antienvironmental regime took over at the national level and in many state governments, the grassroots activists went back to the local level to realize their social and environmental agendas. Riley is convinced that one of the legacies of CRUW and other 1990s grassroots urban environmental groups was to set the stage for the proliferation of urban watershed councils. For example, in Los Angeles sophisticated multiparty watershed groups grew out of the early urban-stream movement. The San Francisco Bay area has about forty of these collaborations. The same is true in the Portland metropolitan region. When Esther Lev and I founded the Urban Streams Council and participated in CRUW's founding, there were no watershed councils in the Portland region. Today, several highly effective watershed councils—all of which embody CRUW's "bottom-up," grassroots philosophy—are engaged in CRUW's core mission and more. This phenomenon has occurred throughout the United States and Canada. The same people who founded CRUW are now busy at the local level creating, leading, and energizing watershed councils.

The Los Angeles River is an excellent case study in changing attitudes toward urban waterways. It can be argued that the Los Angeles River exemplifies the impact that CRUW and its member organizations like the Friends of the Los Angeles River had on elevating the cause of urban stream and river restoration. One of the presenters at the Friends of Trashed Rivers I conference was Lewis McAdams, founder of the Friends of the Los Angeles River, an organization that Riley argues symbolized the marginalization of the urban-river movement by the mainstream environmental community and agencies. The Los Angeles River was something of a joke in those days.

By contrast, in 2003 the Los Angeles River was featured as one of the state's compelling water issues[22] and was featured as the cover story in the highly respected Water Education Foundation's magazine *Western*

Water.[23] In 2001, California's Governor Gray Davis allocated nearly $60 million for the purchase of Los Angeles riverside properties to initiate the creation of more parks along the river. Recently, the *Los Angeles Times* announced that the City of Los Angeles established a $3 million fund to assist in restoration of the Los Angeles River.[24] Finally, a new state agency, the Los Angeles and San Gabriel Rivers Conservancy, has been formed by the California legislature to coordinate and implement restoration of these two urban rivers.[25]

Similar success stories from across the country demonstrate that what most federal and state agencies and even many mainstream environmental organizations thought of as a lost cause is now considered integral to neighborhood and city revitalization efforts. CRUW's founding members and fledgling spinoff "friends" organizations and watershed councils can and should take much of the credit for the transformation of urban waterways as cast-off, disposable natural resources into urban oases or "ribbons of green"—a vision that William H. Whyte championed in his seminal 1968 book, *The Last Landscape.* Whyte's vision for unchannelized, restored streams was captured in *The Last Landscape:* "Nature has laid down a regional design of streams and valleys that provide superb natural connectors, and into the very heart of the urban area. Where streamside land has not been secured against development, it should be; where continuity has been broken, the pieces should be reclaimed wherever it is at all possible.... As a major part of their new beautification programs, cities should launch stream bank improvement projects that save what has not been concreted or rip-rapped; they should also launch projects to bring back to life some of the stretches which have been concreted. Another possibility worth exploring is the unburying of streams."[26]

Organizations like CRUW and individuals like A. L. Riley, Esther Lev, Tom Murdoch, Neil Armingeon, Dennis O'Connor, Karen Firehock, Julie Vincentz, Don Elder, Steve Barnes, and Andrew Moore have in the intervening nearly four decades transformed what Whyte argued might be a "possibility worth exploring" into a reality. Today, thanks to CRUW's pioneering networking and Friends of Trashed Rivers conferences, daylighting streams and urban-stream restoration are now mainstream environmental and social-justice issues. Virtually all of the original CRUW steering committee and board continue to be engaged in urban-stream protection and restoration.[27]

As Laurene Von Klan, founding CRUW board member and director of the Friends of the Chicago River, succinctly put it recently: "In a world where things look pretty bleak for freshwater ecosystems, the Chicago River is getting better all the time. CRUW was a great example of issue-based action. Right place. Right people. All at the right time."[28]

Notes

1. A product of the Oregon Progress Board, the *State of the Environment Report* (2000) provided information on how well the state was protecting Oregon's environment. The Oregon Progress Board is an independent state planning and oversight agency. Created by the legislature in 1989, the Board is responsible for monitoring the state's twenty-year strategic vision, Oregon Shines. The twelve-member panel, chaired by the governor, is made up of citizen leaders and reflects the state's social, ethnic and political diversity. *Available at* http://egov.oregon .gov/DAS/OPB/docs/SOER2000/Ch2_0.pdf (last visited Nov. 8, 2005).

2. Pam Wiley, *No Place for Nature: The Limits of Oregon's Land Use Program in Protecting Fish and Wildlife Habitat in the Willamette Valley*, Defenders of Wildlife (2001), *available at* http://www.biodiversitypartners.org/reports/Wiley/ Recommendations.shtml.

3. Anne Whston Spirn, *The Granite Garden: Urban Nature and Human Design* 5 (1984).

4. Witold Rybczynski, *A Clearing in the Distance: Frederick Law Olmsted and America in the Nineteenth Century* (2000).

5. H. W. S. Cleveland, *Landscape Architecture, as Applied to the Wants of the West* (2002).

6. Karl Haglund, *Inventing the Charles River* (2003).

7. *Report of the Park Board, Portland, Oregon 1903, with the Report of Messrs. Olmsted Bros., Landscape Architects, Outlining a System of Parkways, Boulevards and Parks for the City of Portland* 19–20 (1903).

8. Metro Disappearing Streams Map, Metro Data Resource Center, 600 NE Grand, Portland, OR 97232, *available at* www.metro-region.org (last visited Nov. 8, 2005).

9. Paul Ketcham, Audubon Society of Portland, 1000 Friends of Oregon, *To Save or to Pave: Planning for the Protection of Urban Natural Areas* (Audubon Society of Portland, 1000 Friends of Oregon, 1994). Oregon's statewide land-use program, the first of its kind in the United States, was established by Senate Bill 100 in 1970. The basic premise of the program was that urban-growth boundaries (UGBs) were established around every city in the state. The program was the inspiration of a progressive Republican governor, Tom McCall, with the strong support of the agricultural community. The primary functions of the UGBs were to focus intensified urban development inside the UGBs and to protect commercial agricultural and forestry land outside the UGBs. While one of the program's

nineteen statewide goals, goal 5, was created to address fish and wildlife habitat, wetlands, and other natural resource issues, this "catch-all" goal, until quite recently, has never enjoyed strong political support in the state capital or among urban planners and planning commissions.

10. David Goode, *Wild in London* (1986).

11. L. W. Adams & D. L. Leedy, eds., *Wildlife Conservation in Metropolitan Environments* (National Institute for Urban Wilderness, Columbia, MD 1991).

12. Seven Country in the City symposia were held at Portland State University in the late 1990s and early 2000s. The symposia focused on the protection and restoration of natural resources (streams and wetlands, in particular) and the creation of a bistate regional parks and greenspace system for the Portland-Vancouver metropolitan region.

13. Chris Brown and his colleagues in the Rivers and Trails Conservation Assistance Program (http://www.nps.gov/rtca) were early supporters of the Coalition to Restore Urban Waters. They provided funding to the early Friends of Trashed Rivers conferences for scholarships for low-income and grassroots advocates who otherwise could not afford to attend.

14. Multiobjective management (MOM) is a concept that has since fallen in disuse. It was developed and promulgated by professional floodplain and wetland management professionals within the ranks of the national Association of State Floodplain Managers (www.floods.org) and national Association of State Wetland Managers (www.aswm.org). Multiobjective floodplain and river management holds that the use of nonstructural, nonengineered management alternatives that recognize the multiple values of riverine systems yield greater social, ecological, and economic benefits than single-purpose engineered solutions such as dams and levees. MOM reached its nadir during the Midwest floods in the mid-1990s. MOM was a powerful concept that was readily adopted by grassroots stream-restoration advocates, given MOM's validation of appropriate technology, nonstructural approaches (including soil bioengineering and other low-tech, hands-on restoration techniques).

15. Stream daylighting involves digging up streams and liberating them from their underground culverts. In the case of Strawberry Creek, approximately 300 feet of stream was daylighted, and its culvert was used to introduce topographic relief at the site. The result was a new park that now occupies a formerly high-drug-use abandoned rail yard. The Waterways Restoration Institute shares offices with architectural and other professional services and adjoins a bakery. The nearby school utilizes the park for environmental education programs, and drug use is now nonexistent at the site.

16. Neil Armingeon, Lake Pontchartrain Basin Foundation; Steve Barnes, New York/New Jersey Baykeeers; Jack Byrne; Karen Firehock, Isaak Walton League's Save Our Streams program; Mike Houck, Portland Audubon Society, Portland, OR; Esther Lev, Urban Streams Council of the Wetlands Conservancy, Tualatin, OR; Andrew Moore, National Association of Service and Conservation Corps; Tom Murdoch, Adopt-a-Stream Foundation, Seattle, WA; Laurence Von Klan, Friends of the Chicago River.

17. A Bill to Amend the Watershed Protection and Flood Prevention Act to Establish a Waterways Restoration Program, H.R. 1331, 104th Congr., 1st Sess. (1996). This legislation contained the following language: "restoring degraded streams, rivers, and other waterways to their natural state is a cost-effective means to control flooding, excessive erosion, sedimentation, and nonpoint pollution, including stormwater runoff; low-income and minority communities frequently experience disproportionately severe degradation of waterways in their communities but historically have had difficulty in meeting eligibility requirements for Federal watershed projects."

18. Coalition to Restore Urban Waters Long Range Plan (Aug. 1995). Julie Vincentz, CRUW coordinator, Izak Walton League of America, Save Our Streams Program, 707 Conservation Lane, Gaithersburg, MD 20878-2983.

19. The Steering Committee that helped create most of CRUW's programs, plan Friends of Trashed Rivers conferences, and implement CRUW's mission included Deborah Alex-Saunders, Minority Environmental Association, Sandusky, OH; Neil Armingeon, Lake Ponchartrain Basin Foundation, Metairie, LA; Steve Barnes, NY/NJ Baykeeper, Highlands, NJ; Jack Byrne, River Watch Network, Montpelier, VT; Karen Firehock, Izaak Walton League of America, Save Our Streams Program, Gaithersburg, MD; Mike Houck, Audubon Society of Portland, Portland, OR; Elbert Jenkins, Minority Environmental Association, Decatur, GA; Esther Lev, Urban Streams Council, Tualatin, OR; Andrew Moore, National Association of Service and Conservation Corps, Washington, DC; Tom Murdoch, Adopt-a-Stream Foundation, Everett, WA; Dennis O'Connor, Restoration Ecologist, Portland, OR; A. L. Riley, Waterways Restoration Institute, Berkeley, CA; Bruce Saito, Los Angeles Conservation Corps, Los Angeles, CA; Laurence Von Klan, Friends of the Chicago River, Chicago, IL.

20. Chair, Neil Armingeon; vice chair, Tom Murdoch; treasurer, Laurence Von Klan; secretary, Julie Vincentz; Board, Cesar Avila, Esther Lev, Mike Houck, Steve Barnes, Andy Moore, Jack Byrne.

21. A. L. Riley, *Restoring Streams in Cities: A Guide for Planners, Policymakers, and Citizens* (1998).

22. Personal communication, A. L. Riley (May 12, 2005).

23. Water Education Foundation, Western Water magazine, Nov./Dec. 2003, *available at* www.water-ed.org/novdec03.asp (last visited Nov. 8, 2005).

24. Jessica Garrison, *River Plan Rolls Along: Efforts to Revitalize L.A.'s Concrete-Lined Waterway Will Get a Boost as City Leaders Begin a Search for a Design Consultant*, Los Angeles Times, Nov. 4, 2004, at B1 ("There are those such as Councilman Ed Reyes who view revitalization as a path to an utopian mix of parks, natural habitat, affordable housing and commercial development. The councilman even talks of donning swim trunks and frolicking in the waters with his children").

25. Personal communication, A. L. Riley (May 12, 2005).

26. William H. Whyte, *The Last Landscape* 351 (2002) (1968).

27. Neil Armingeon turned in his Mardi Gras beads, left New Orleans, and is, in his own words, "up to his butt in developers and bad land-use planning." He's

living in Jacksonville, Florida, as the St. Johns Riverkeeper (www.stjohnsriver keeper) in the first and fourth fastest-growing counties in the country, where "they are fighting the good fight and winning a few." Don Elder is still with River Network and in April 2003 became its president. It is still doing the national River Rally, which Don says has grown from the initial 100 attendees to more than 500 (www.rivernetwork.org/rally). Mike Houck founded his own nonprofit in 1999, the Urban Greenspaces Institute, whose motto is "In Livable Cities Is Preservation of the Wild." The Institute is housed at Portland State University's Center for Spatial Analysis and Research in the Geography Department, where he is an adjunct instructor. He continues to work part-time at Portland Audubon Society (www.audubonportland.org). Andrew Moore consults with mayors and other city officials to provide more work, education, and training opportunities for disconnected youth through such transitional job programs as conservation corps and YouthBuild. Laurene Von Klan moved on from her director position at the Friends of the Chicago River (www.chicagoriver.org) to her new position as president and CEO of the Chicago Academy of Sciences and Peggy Notebaert Nature Museum. Esther Lev, cofounder of the Urban Streams Council, is now Executive Director of the Wetlands Conservancy (www.wetlandsconservancy. org), based in Tualatin, Oregon. Esther has assumed a leadership role in wetland conservation and expanded the Conservancy's wetland conservation efforts statewide throughout Oregon. She currently serves on the regional government's (Metro) Greenspaces Policy Advisory Committee. Tom Murdoch is celebrating his twentieth anniversary with the Adopt-a-Stream Foundation (www.stream keeper.org) and is still fundraising for his twenty-acre Northwest Stream Center. He's still running his Streamkeeper Academy (www.streamkeeper.org). Dennis O'Connor is a private consultant restoring urban streams in the Portland metro-politan region and elsewhere around the country. Dennis also volunteers his stream-restoration expertise to groups like Portland Audubon Society. A. L. Riley is working for the state of California's Water Resources Board and continues to volunteer with the Urban Creeks Council and Waterways Restoration Institute. Julie Vincentz left the Izaak Walton League in 2000 to seek a master's degree in public administration. She is now working at the Congressional Budget Office (CBO) as a budget analyst, where she continues to be involved in water-related issues through her work analyzing the budgets of the Bureau of Reclamation, the Army Corps of Engineers, and FEMA/Flood Insurance.

28. Personal communication (April 2005).

9

Bankside Katrina: A Postscript

Paul Stanton Kibel

The essays solicited for *Rivertown* were completed before Hurricane Katrina's storm surge pushed over and through New Orleans' levees. By the time the hurricane hit in August 2005, the book was already in its final stages, but given Katrina's human and ecological implications for the national urban-river debate, it was not possible to ignore the event. Melissa Samet revised chapter 7 to include some analysis of New Orleans–specific concerns, but the magnitude of the disaster warrants some further discussion.

From an environmental standpoint, Katrina raises many complex questions. In terms of environmental justice, why did New Orleans' low-income, African American citizens bear the brunt of the flooding and death? In terms of global warming, to what extent did the Gulf of Mexico's rising water temperatures, a by-product of regional oil refineries, create climatic conditions that increased the hurricane's severity? In terms of land use, how did wetlands loss and canal construction contribute to the storm surge that inundated New Orleans? These critical questions merit close scrutiny, but the last question bears most directly on *Rivertown*'s subject matter because it focuses our attention on the transformation of lands that front and surround urban waterways.

The struggle to make sense of Katrina is the latest chapter in a much older story—the tale of New Orleans' response to its inhospitable natural surroundings. Bordered by the Mississippi River to the west, Lake Pontchartrain to the east, and the volatile Gulf waters nearby to the south, the city has faced flooding problems since its founding. The primary response to this flood threat for more than two centuries has been to build ever-higher levees along ever-increasing stretches of the New Orleans waterfronts. This response was soon seen to be problematic from a long-term engineering standpoint for two reasons: (1) the levees constructed to protect New Orleans from inundation also restricted the surrounding waters, causing water levels outside the city to rise and increasing the risk of flooding, and (2) when Lake Pontchartrain or the Mississippi River topped the levees and entered New Orleans, the levees acted to collect and retain waters in the city, worsening and prolonging flood conditions.

The shortcomings to levy-based flood protection for New Orleans have been long known. In 1846, the state engineer of Louisiana, P. O. Herbert, observed: "Every day, levees are extended higher and higher up the river— natural outlets closed, and every day the danger to the city of New Orleans and to all the lower country in increased. . . . the gradual elevation of the bed of the river is the inevitable consequence of confining its turbulent waters between levees. This operation of course increases the danger of inundation."[1] Similarly, in 1858 the state engineer of Louisiana, A. D. Wooldridge, warned: "I find myself forced to the conclusion that the entire dependence on the levee system is not only unsafe for us, but I think will be destructive to those who shall come after us."[2]

One hundred fifty years ago, Herbert and Wooldridge advocated for more outlets upstream of New Orleans so that rising Mississippi River waters could spread out into the floodplain a safe distance from the city. The upstream rural planters, however, built their own levees to protect their farmlands from the very inundation that would lessen flood hazards in New Orleans. State and federal authorities lacked the political resources and will to push through the local land-use changes necessary to implement the outlet alternative. As Craig E. Colten, professor of geography at Louisiana State University and author of the 2005 book *An Unnatural Metropolis: Wresting New Orleans from Nature*, explains: "Relieving storm-induced urban flooding was a secondary issue. Only as the city achieved the primary objective [levee protection from Mississippi River flooding] did the secondary concern become obvious. Simply put, once the bowl was securely in place, rain [from summer storms and hurricanes] began to fill it. Protection against one hazard guaranteed the city's long-standing struggle with the second."[3]

In the twentieth century, New Orleans once again turned to levees rather than land-use management to deal with flood control. The marshes between New Orleans and the Gulf of Mexico traditionally served to absorb and dilute hurricane-driven storm surges, usually rendering them benign by the time they reached the city. In recent decades, however, this has changed.

First, the coastal wetlands between New Orleans and the Gulf of Mexico are being lost. This loss has many causes, including the absence of sediment replenishment (because the levees prevent natural river flooding), draining ("reclamation" in policy parlance) to facilitate the construction of residences and resorts near the Gulf coast, and the excavation of water channels to facilitate oil and gas development.[4] The effects of these wetlands-adverse activities on hurricane storm surges headed toward New Orleans is well documented.

In terms of marsh loss, Louisiana has lost 1.2 million acres of coastal wetlands.[5] As Carol Browner, former administrator of the United States Environmental Protection Agency, explained: "The wetlands act like a sponge in a storm. They're an incredibly smart and helpful part of nature. But they have to be kept moist, like a sponge on your kitchen counter. If they're dried out and developed, they don't work."[6] Browner's point was echoed in a *New York Times* article on Hurricane Katrina: "Marshes that once absorbed storms have been allowed to die off and shrink, leaving stretches of open water that can be flung shoreward by storm surges."[7] *Urban Land* magazine, in its special January 2006 edition on rebuilding the Gulf Coast, noted: "The major cause of Katrina's extreme devastation to the city of New Orleans is that the wetlands act as a speed bump for storm surges caused by hurricanes. . . . If the current rate of wetlands erosion continues, New Orleans will be left increasingly vulnerable to the potential storm surge brought by every approaching hurricane season."[8]

Second, a navigation channel constructed by the Army Corps in the 1960s now amplifies hurricane storm surges that are making their way from the Gulf of Mexico toward New Orleans. As the *Washington Post* reported: "Today, exactly eight weeks after [Katrina], all three breaches are looking less like acts of God and more like failures of engineering that could have been anticipated and very likely prevented. . . . In 1965, the Corps completed the 76-mile long, 36-foot-deep Mississippi River Gulf Outlet. . . . The outlet, known locally as MRGO or 'Mr. Go,' created a navigation shortcut to the Port of New Orleans, although a little-used one that averages fewer than one ship a day. But the outlet also amounted to a funnel that would accelerate and enlarge any storm surges headed for the city's levees."[9]

Three months before Katrina hit, Hassan Mashriqui of Louisiana State University's Hurricane Center warned a gathering of emergency managers that the MRGO was a "Trojan horse" for hurricanes to ride into New Orleans and that the outlet could amplify storm surges 20 to 40 percent.[10] Even the conservative *National Review* published a post-Katrina article critical of the MRGO, noting, "After the Katrina disaster, many in the press criticized the Bush Administration for underfunding the Army Corps of Engineers. These critics missed the point. The problem is not under-funding but the total lack of prioritization that characterizes the Corps' activities.... Shortly after the Corps completed the MRGO [Mississippi River Gulf Outlet] in 1965, Hurricane Betsy hit New Orleans and sent a storm surge up the channel into St. Bernard's parish, resulting in massive flooding. Betsy killed more than 70 people and caused over $1 billion worth of damage."[11] According to the *National Review*, Katrina is evidence that the lessons of Betsy were not learned.

In the aftermath of Katrina, there have been efforts to redirect criticism away from the engineering defects of the levee-based flood-protection system, wetlands loss, and MRGO and to refocus criticism on environmentalists and environmental laws. One such effort was made by Michael Tremoglie of *Front Page* magazine in an article entitled *New Orleans: A Green Genocide*. Tremoglie states: "As radical environmentalists continue to blame the ferocity of Hurricane Katrina's devastation on President Bush's ecological policies, a mainstream Louisiana media outlet inadvertently disclosed a shocking fact: Environmental activists were responsible for spiking a plan that may have saved New Orleans.... Decades ago, the Green left, pursuing its agenda of valuing wetlands and topographical 'diversity' over human life, sued to prevent the Army Corps of Engineers from building floodgates that would have prevented significant flooding that resulted from Hurricane Katrina.... Despite its pious rhetoric, the environmental left's true legacy will be on display in New Orleans for years to come."[12]

Conservatives, including former chair of the House of Representatives' Resources Committee, Richard Pombo of California, seized on the 1977 litigation referenced in Tremoglie's *Front Page* article to make the case that the National Environmental Policy Act (NEPA)—the law at issue in the lawsuit—needs to be reformed. More specifically, Pombo has played a key role in the formation of the NEPA Task Force, which, among other things, suggested that navigation projects (such as Army Corps levee and channel work near New Orleans) and energy projects (such as new oil and gas development in coastal Louisiana) should be exempt from NEPA's requirement that an environmental impact statement (EIS) be prepared prior to federal project approvals. On September 9, 2005, Pombo's NEPA Task Force announced that it would be holding hearings to find ways to "cut through the red tape so that what we experienced in the wake of Katrina does not happen again."[13]

Whether or not the efforts of Tremoglie and Pombo to use Katrina to limit the scope of NEPA will find political traction remains to be seen. However, the analysis that underlies their efforts has been challenged on the grounds that it disregards the reasons that the floodgate project challenged in the 1977 litigation was ultimately rejected. According to Douglas Kysar, professor of environmental law at Cornell University School of Law, the project referenced in the Tremoglie article was later abandoned in large part because of evidence that the project itself would cause significant damage to wetlands that were needed to absorb storm surges.[14] Since the proposed floodgates project itself was later passed over by the federal government to preserve wetlands that it determined were needed to dissipate storm surges, Tremoglie's claim that project opponents valued wetlands over human life emerges as somewhat nonsensical. Additionally, had an environmental impact statement been prepared for the Mississippi River Gulf Outlet in the 1960s, the storm-surge-inducing effect of the project might have been more fully understood, and the outlet canal might not have been built. Unfortunately, NEPA did not go into effect until 1970. The experience with the MRGO, however, makes a strong case for the role NEPA's impact-assessment requirements can play in identifying and avoiding projects that increase flood hazards.

The *National Review*, although critical of the MRGO, has also pointed to environmentalists and environmental laws as contributing to Katrina's inundation of New Orleans. More specifically, in the September 8, 2005, issue of the magazine, national environmental groups were criticized for bringing a NEPA lawsuit in 1996 that alleged a failure to conduct adequate environmental assessment of a proposal by the United States Army Corps of Engineers to raise and fortify 300 miles of levees along the

Mississippi River.[15] The article in the *National Review* stated that "Whether this delay directly affected the levees that broke in New Orleans is difficult to ascertain."[16] As it turns out, however, this is not difficult to ascertain because the breaks in the levees holding back Lake Pontchartrain (on the east side of the city) rather than those on the Mississippi River (on the west side of the city) resulted in the inundation of New Orleans during Katrina.[17] These reporting errors aside, Katrina has prompted a vigorous political debate about how to best deal with flood-control considerations in the urban context, and the outcome of this debate has implications not only for New Orleans but for riverside cities around the country.

Although the bulk of post-Katrina attention has been on levee restoration, there are some indications that the wetlands dimension has not been completely overlooked. For instance, the proposed Davis Pond diversion project calls for the release of up to 10,650 cubic feet per second of freshwater into a 33,000 acre marshland basin.[18] This release is intended to imitate the inundation of nutrients and sediment that used to occur naturally with spring floods. More projects like Davis Pond are needed to reestablish the ecological speed bumps to slow down and dissipate gulf storms.

The essays in Rivertown can help contribute to an informed post-Katrina debate over urban waterways and urban flooding hazards. The current responses to Katrina call to mind chapter 2 by Robert Gottlieb and Andrea Misako Azuma on the Los Angeles River and chapter 8 by Mike Houck on the work of the Coalition to Restore Urban Waterways. These essays reveal that levees and armoring are not the only means of preventing urban rivers from flooding and that alternative approaches can often prove more effective, less expensive, and more compatible with local economic and environmental needs.[19] These lessons must not be lost in Katrina's wake.

Notes

1. Craig E. Colten, *An Unnatural Metropolis: Wresting New Orleans From Nature* 25–26 (2005).
2. *Id.* at 29.
3. *Id.* at 142.

4. *Lawsuit Filed against Major Oil Companies Alleging Ecological Damage Worsened Katrina*, Environmental News Network, Sept. 26, 2005 ("the lawsuit, filed in the United States District Court for the Eastern District of Louisiana, alleged that the major oil companies' oil, gas and pipeline exploration and drilling activities throughout Southeast Louisiana resulted in ecological damages to such an extent that coastal marshes were destroyed which previously had protected New Orleans naturally from Katrina level hurricane wind and tidal surges"). *See also* the following information provided by the Louisiana Sea Grant College Program *available at* http://www.laseagrant.org/hurricane/wetlands.htm: "As the levees grew larger, the 'wild' nature of the river was restricted. This ultimately reduced the frequency of alluvial flooding and new delta lobe formation that is so critical to the creation and maintenance of wetlands in coastal Louisiana.... The dredging of thousands of miles of access canals for petroleum extraction and navigation has accelerated saltwater intrusion."

5. Posting on the Web site of the Louisiana Sea Grant College Program (Oct. 4, 2005).

6. Jane Mayer, *Wind on Capitol Hill*, The New Yorker, Sept. 19, 2005, *available at* http://newyorker.com/talk/content/article/05919ta_talk_mayer (last visited Mar. 6, 2006).

7. Donald G. McNeil Jr., *Imagine Twenty Years of This*, N.Y. Times, (§ 4), Sept. 25, 2005, at 4.

8. Charles Picciola, *Louisiana's Coastal Plight*, Urban Land, Jan. 2006, at 70.

9. Joby Warwick and Michael Grunwald, *Investigators Link Levee Failures to Design Flaws*, Washington Post, Oct. 24, 2005, at A1.

10. *Id.*

11. Stephen Spruiell, *Soft Corps*, National Review, Oct. 10, 2005, *available at* http://nationalreview.com/comment/spruiell200509231329.asp (last visited Mar. 6, 2006).

12. Michael Tremoglie, *New Orleans: A Green Genocide*, Front Page Magazine, Sept. 8, 2005, *available at* http://frontpagemag.com/Articles/Read/Article.asp?ID=19418 (last visited Sept. 12, 2005).

13. *Katrina Recovery Become Rallying Cry for NEPA Reform Advocates*, Inside Washington Publishers, Sept. 9, 2005.

14. Ralph Vartabedian and Richard B. Schmidt, *Katrina Has Role in Fight over Environmental Law*, Seattle Times, Sept. 19, 2005, *available at* http://www.seattletimes.nwsource.com/html/nationworld/2002504609_katcorps19.html (last visited Mar. 6, 2006).

15. Jerry Mitchell, *Emails Suggest Government Seeking to Blame Group*, Clarion-Ledger, Sept. 16, 2005, *available at* http://www.clarionledger.com/apps/pbcs.dll/article?AID=/20050916/NEWS0110/5091603 (last visited Mar. 6, 2006).

16. *Id.*

17. *Id.*

18. Picciola, *supra* note 8, at 71.

19. Michael Grunwald, *Par for the Corps: A Flood of Bad Projects*, Washington Post, May 14, 2006, at B01 ("Somehow, America has concluded that the scandal of Katrina was the government's response to the disaster, not the government's contribution to the disaster. The Corps has eluded the public's outrage—even though a useless Corps shipping canal intensified Katrina's surge, even though poorly designed Corps floodwalls collapsed just a few feet from an unnecessary $750 million Corps navigation project, even though the Corps had promoted development in dangerously low-lying New Orleans floodplains and had helped destroy the vast marshes that once provided the city's natural flood protection.")

Bibliography

This is a partial list of the books that are discussed and cited in this book or that otherwise influenced and informed it.

Brechin, Gray, *Imperial San Francisco: Urban Power, Earthly Ruin* (University of California Press 1999).

Breen, B.A., et al., *Waterfronts: Cities Reclaim Their Edge* (Waterfront Press 1997).

Colten, Craig E., *An Unnatural Metropolis: Wresting New Orleans from Nature* (Louisiana State University Press 2005).

Davis, Mike, *City of Quartz: Excavating the Future in Los Angeles* (Verso 1990).

Gandy, Matthew, *Concrete and Clay: Reworking Nature in New York City* (MIT Press 2002).

Goode, David, *Wild in London* (Michael Joseph Ltd. 1986).

Gordon, David, *Green Cities: Ecologically Sound Approaches to Urban Space* (Black Rose 1989).

Gumprecht, Blake, *The Los Angeles River: Its Life, Death and Possible Rebirth* (Johns Hopkins University Press 1999).

Haglund, Karl, *Inventing the Charles River* (MIT Press 2003).

Hill, Libby, *The Chicago River: A Natural and Unnatural History* (Lake Claremont Press 2000).

Houck, Michael C., et al., *Wild in the City: A Guide to Portland's Natural Areas* (Oregon Historical Society 2000).

Jacobs, Jane, *The Death and Life of Great American Cities* (Random House 1961).

Jensen, Jens, *Siftings* (Johns Hopkins University Press 1990) (1939).

Moe, Richard, & Curter Wilkie, *Changing Places: Rebuilding Community in the Age of Sprawl* (Henry Holt 1997).

Mumford, Lewis, *The City in History: Its Origins, Its Transformation and Its Prospects* (Harcourt, Brace and World 1961).

Orsi, Jared, *Hazardous Metropolis: Flooding and Urban Ecology in Los Angeles* (University of California Press 2004).

Otto, Beth, Kathleen McCormick & Michael Leccese, *Ecological Riverfront Design: Restoring Rivers, Connecting Communities* (American Planning Association 2004).

Petts, G., J. Heathcote & D. Martin, eds., *Urban Rivers: Our Inheritance and Future* (IWA 2002).

Pyle, Robert Michael, *The Thunder Tree: Lessons from an Urban Wildland* (Lyons Press 1998).

Riley, Ann L., *Restoring Streams in Cities: A Guide for Planners, Policymakers and Citizens* (Island Press 1998).

Rybczynski, Witold, *A Clearing in the Distance: Frederick Olmsted and America in the Nineteenth Century* (Touchstone 2000).

Spirn, Anne W., *The Granite Garden: Urban Nature and Human Design* (Basic Books 1984).

Tarr, Joel A., *The Search for the Ultimate Sink: Urban Pollution in Historical Perspective* (University of Akron Press 1996).

Contributors

Andrea Misako Azuma currently serves as a project manager at the Center for Food and Justice at the Urban and Environmental Policy Institute of Occidental College. She was project manager for the Re-Envisioning the Los Angeles River project and holds a bachelor's degree from Occidental College and a M.S. degree from Cornell University.

Uwe Steven Brandes managed the planning effort known as the Anacostia Waterfront Initiative from 2000 to 2004 at the District of Columbia Office of Planning. He currently serves as strategic adviser to the newly formed Anacostia Waterfront Corporation.

Robert Gottlieb is the Henry R. Luce Professor of Urban and Environmental Policy at Occidental College in Los Angeles. He also directs the Urban and Environmental Policy Institute (UEPI) at Occidental. Gottlieb has written ten books, including *The Next Los Angeles: The Struggle for a Livable City* (coauthored with UEPI colleagues Mark Vallianatos, Regina M. Freer, and Peter Dreier) (2005). He was the principal investigator for the Re-Envisioning the Los Angeles River project that is described in chapter 2. Financial support for the Re-Envisioning program came from the California Council for the Humanities.

Mike Houck is the founder and director the Urban Greenspaces Institute (which operates out of the Department of Geography at Portland State University in Oregon) and has been the Audubon Society of Portland's urban naturalist since 1982, when the Society initiated its Urban Naturalist Program. With M. J. Cody, he also coedited the book *Wild in the City: A Guide to Portland's Natural Areas* (2000).

Paul Stanton Kibel is an adjunct professor of environmental law and directs the City Parks Project at Golden Gate University School of Law in San Francisco. He teaches water policy at Berkeley's Goldman School of Public Policy. He is also of counsel to and a former partner with Fitzgerald Abbott & Beardsley and director of Policy West—a public-policy consultancy in Alameda, California. His previous publications include the book *The Earth on Trial: Environmental Law on the International Stage* (1999). He holds a LL.M. from Berkeley's Boalt Hall Law School and a B.A. from Colgate University.

Ron Love holds bachelor degrees in mathematics (University of Texas at El Paso) and urban planning (University of Utah) and a master's of public administration from the University of Utah. He retired as a major from the U.S. Army Corps of Engineers and currently is a technical planner in the Public Services (Public Works) Department of the Salt Lake City Corporation. He recognizes the tremendous assistance of Scott Stoddard—civil and environmental engineer, intermountain representative of the Sacramento District, U.S. Army Corps of Engineers—in researching, drafting, and revising this essay.

Richard Roos-Collins is a senior attorney with the Natural Heritage Institute (NHI) in San Francisco. He served as lead counsel for the Guadalupe-Coyote Resource Conservation District in the litigation and negotiation of the matters discussed in Chapter 6. He has a J.D. from Harvard Law School (1986) and a B.A. in English from Princeton University (1975). Julie Gantenbein, NHI staff attorney, assisted him in preparation of his essay.

Melissa Samet directs American Rivers' Army Corps Reform Campaign. She joined American Rivers in February 2001 after working for six years at Earthjustice Legal Defense Fund. Samet coauthored the report *Costly Corps: How the U.S. Army Corps of Engineers Uses Your Tax Dollars to Destroy the Environment* (1996). She received her J.D. from the New York University School of Law and holds a B.S. in wildlife biology from the University of Vermont.

Christopher Theriot is a lecturer in metropolitan environmental problems at Roosevelt University in Chicago and formerly worked with the City of Chicago's Environmental Department. He holds a master's in public policy from the University of Chicago and a bachelor of arts from Colgate University. Pete Mulvaney collaborated in developing the concept for Chapter 4.

Kelly Tzoumis is a professor and director of public policy studies at DePaul University in Chicago. She has a B.S. and master's of public administration from Iowa State University and a Ph.D. from Texas A&M University. She has written several articles on environmental policy and a book, *Environmental Policy Making in Congress from 1789–1999: Issue Definitions of Wetlands, the Great Lakes, and Wildlife Policies* (2001).

Credits

A previous version of chapter 1, *Bankside Urban: An Introduction*, by Paul Stanton Kibel, was published in the 2005 City Rivers symposium edition of the *Golden Gate University Law Review* (vol. 35, no. 3) under the title *An Introduction to the Issue: The Urban Bankside*. It is republished by special permission of *Golden Gate University Law Review*.

A previous version of chapter 2, *Bankside Los Angeles*, by Robert Gottlieb and Andrea Misako Azuma, was published in the 2005 City Rivers symposium edition of the *Golden Gate University Law Review* under the title *Re-Envisioning the Los Angeles River: An NGO and Academic Institution Influence the Policy Discourse*. It is republished by special permission of *Golden Gate University Law Review*.

A previous version of chapter 3, *Bankside Washington, D.C.*, by Uwe Steven Brandes, was published in the 2005 City Rivers symposium edition of the *Golden Gate University Law Review* under the title *Recapturing the Anacostia River: The Center for Twenty-first Century Washington, D.C.* It is republished by special permission of *Golden Gate University Law Review*.

A previous version of chapter 4, *Bankside Chicago*, by Christopher Theriot and Kelly Tzoumis, was published in the 2005 City Rivers symposium edition of the *Golden Gate University Law Review* under the title *Deep Tunnels and Fried Fish: Tracing the Legacy of Human Intervention on the Chicago River*. It is republished by special permission of *Golden Gate University Law Review*.

A previous version of chapter 5, *Bankside Salt Lake City*, by Ron Love, was published in the 2005 City Rivers symposium edition of the *Golden Gate University Law Review* under the title *Daylighting Salt Lake's City Creek: An Urban River Unentombed*. It is republished by special permission of *Golden Gate University Law Review*.

A previous version of chapter 6, *Bankside San Jose*, by Richard Roos-Collins, was published in the 2005 City Rivers symposium edition of the *Golden Gate University Law Review* under the title *A Perpetual Experiment to Restore and Manage Silicon Valley's Guadalupe River*. It is republished by special permission of *Golden Gate University Law Review*.

Index

Urban and Industrial Environments
Series editor: Robert Gottlieb, Henry R. Luce Professor of Urban
and Environmental Policy, Occidental College

Brian K. Obach, *Labor and the Environmental Movement: The Quest for Common Ground*

Peggy F. Barlett and Geoffrey W. Chase, eds., *Sustainability on Campus: Stories and Strategies for Change*

Steve Lerner, *Diamond: A Struggle for Environmental Justice in Louisiana's Chemical Corridor*

Jason Corburn, *Street Science: Community Knowledge and Environmental Health Justice*

Peggy F. Barlett, ed., *Urban Place: Reconnecting with the Natural World*

David Naguib Pellow and Robert J. Brulle, eds., *Power, Justice, and the Environment: A Critical Appraisal of the Environmental Justice Movement*

Eran Ben-Joseph, *The Code of the City: Standards and the Hidden Language of Place Making*

Nancy J. Myers and Carolyn Raffensperger, eds., *Precautionary Tools for Reshaping Environmental Policy*

Kelly Sims Gallagher, *China Shifts Gears: Automakers, Oil, Pollution, and Development*

Kerry H. Whiteside, *Precautionary Politics: Principle and Practice in Confronting Environmental Risk*

Ronald Sandler and Phaedra C. Pezzullo, eds., *Environmental Justice and Environmentalism: The Social Justice Challenge to the Environmental Movement*

Julie Sze, *Noxious New York: The Racial Politics of Urban Health and Environmental Justice*

Robert D. Bullard, ed., *Growing Smarter: Achieving Livable Communities, Environmental Justice, and Regional Equity*

Ann Rappaport and Sarah Hammond Creighton, *Degrees That Matter: Climate Change and the University*

Michael Egan, *Barry Commoner and the Science of Survival: The Remaking of American Environmentalism*